FOOD INDUSTRY 4.0

W0193389

FSC
www.fsc.org
MIX
Paper from
responsible sources
FSC® C022174

IFIS

IFIS is a leading international provider of information in the sciences of food and health, and produces FSTA® – Food Science and Technology Abstracts. FSTA is a specialist database of curated content which is indexed using the unique FSTA Thesaurus, the most comprehensive food and beverage thesaurus in the world. As an educational charity, IFIS provides resources and training in information research skills, and provides free access to FSTA in developing countries through Research4Life. IFIS' resources enable researchers, industry practitioners and students in the global food community to find the relevant, reliable research they need. Website: www.ifis.org

**food and health
information**

FOOD INDUSTRY 4.0

Unlocking Advancement Opportunities in the Food Manufacturing Sector

Wayne Martindale

National Centre for Food Manufacturing, University of Lincoln, UK

Linh Duong

University of the West of England, Bristol, UK

and

Sandeep Jagtap

Sustainable Manufacturing Systems Centre, Cranfield University, UK

CABI is a trading name of CAB International

CABI
Nosworthy Way
Wallingford
Oxfordshire OX10 8DE
UK

CABI
WeWork
One Lincoln St
24th Floor
Boston, MA 02111
USA

Tel: +44 (0)1491 832111
E-mail: info@cabi.org
Website: www.cabi.org

Tel: +1 (617)682-9015
E-mail: cabi-nao@cabi.org

A catalogue record for this book is available from the British Library, London.

Library of Congress Cataloging-in-Publication Data

Names: Martindale, Wayne, 1965- author. | Duong, Linh Nguyen Khanh, 1985-
   author. | Jagtap, Sandeep, author.
Title: Food industry 4.0 : unlocking advancement opportunities in the food manufacturing
   sector / Wayne Martindale, Linh Duong and Sandeep Jagtap
Other titles: Food industry four point zero
Description: Oxfordshire ; Boston : CABI, [2022] | Includes bibliographical references and index. |
   Summary: "Highlights the advancement opportunities for the food manufacturing sector, including
   innovation in products, processes and services, and considers disruptive business models which offer
   alternative solutions to the food sector as it seeks to combine productive, efficient and sustainable
   practices. Draws on authors' extensive industry and academic expertise"-- Provided by publisher.
Identifiers: LCCN 2021033222 (print) | LCCN 2021033223 (ebook) | ISBN 9781789248425 (paperback) |
   ISBN 9781789248586 (ebook) | ISBN 9781789248593 (epub)
Subjects: LCSH: Food industry and trade--Production control. | Food
   industry and trade--Technological innovations. | Food supply. | Industry 4.0.
Classification: LCC TP372.7 .M37 2022 (print) | LCC TP372.7 (ebook) | DDC
   664/.024--dc23/eng/20211109
LC record available at https://lccn.loc.gov/2021033222
LC ebook record available at https://lccn.loc.gov/2021033223

References to Internet websites (URLs) were accurate at the time of writing.

ISBN: 9781800621039 (hardback)
      9781789248425 (paperback)
      9781789248586 (ePDF)
      9781789248593 (ePub)

DOI: 10.1079/9781789248593.0000

Commissioning Editor: Rebecca Stubbs
Editorial Assistant: Kate Hill
Production Editor: Tim Kapp

Typeset by SPi, Pondicherry, India
Printed and bound in the UK by Severn, Gloucester

Contents

About the Authors

Wayne Martindale, FIFST, has delivered sustainability research for over 20 years with some of our best-known agri-food brands. He started his career in technical communications and external affairs at Levington Agriculture and has worked at CSIRO Australia, the EC Joint Research Centre and the OECD in Paris across innovation, science and technology. He trained as a management apprentice with British Sugar plc, Bush Boake Allen Group and Ford Motor Company Ltd in the 1980s. His doctorate in biochemistry was awarded in the early 1990s at the University of Sheffield. He now leads a portfolio of research and projects that apply geospatial science, connecting industries to sustainability and measuring the impact of production and consumption.

Food Insights and Sustainability, National Centre for Food Manufacturing, University of Lincoln, UK. E-mail: wmartindale@lincoln.ac.uk

Linh Duong is a senior lecturer in operations management at Bristol Business School, University of the West of England, UK. His current research interests focus on sustainable and resilient supply chain management with the link to digital transformation, innovation and collaboration among supply chain partners. He focuses on vulnerable contexts such as the agri-food industry, tourism industry or small and medium enterprises (SMEs). His papers on supply chain resiliency and sustainable innovation were published in the *International Journal of Production Research* (ABS: 3, JIF: 3.199), *Journal of Macromarketing* (ABS: 2, JIF: 1.952) and *Trends in Food Science and Technology* (JIF: 11.077). He has previously worked at the University of Lincoln, UK, New Zealand Forest Research Institute (Scion) and Auckland University of Technology (New Zealand). From 2007, he worked in supply chain management for dairy and pharmaceutical companies. He also joined projects on distribution management systems, inventory management and production management. Dr Duong has taught a range of operations management and supply chain modules and has experience in module design and student supervision. He is on the Editorial Review Board for the *International Journal of Applied Logistics* and reviews for several journals, including the *International Journal of Production Economics*.

Faculty of Business and Law, University of the West of England, UK. E-mail: linh.duong@uwe.ac.uk

Sandeep Jagtap is a lecturer in smart and green manufacturing at the Sustainable Manufacturing Systems Centre, School of Aerospace, Transport and Manufacturing, Cranfield University. He worked as a lecturer in food and drink supply chain management at the National Centre for Food Manufacturing, University of Lincoln, UK. He has over 15 years' experience within academia and industry. Sandeep holds a PhD in sustainable food supply chain management from Loughborough University, UK, which was sponsored by the EPSRC Centre for Innovative Manufacturing in Food. He holds a BTech in food technology from North Maharashtra University, India, a Master's degree in bio-food technology from Lund University, Sweden, and an MBA from the University of Applied Sciences in Stralsund, Germany. He serves on the Editorial Advisory Board for the *British Food Journal*. He is a Fellow of the Institute of Food Science and Technology (FIFST) and Higher Education Academy (FHEA).

Sustainable Manufacturing Systems Centre, Cranfield University, UK. E-mail: s.z.Jagtap@cranfield.ac.uk

Preface

Research carried out by the authors and presented in this book has opened many areas of work that identify how the manufacturer can provide both security and sustainability in our current Food Industry 4.0 to 5.0 transition (Martindale, 2015). This realization started the route to this book, which seeks to re-balance the positioning of research in food manufacturing. I believe that industrialized food supply chains and, in particular, food manufacturers are very much part of a solution to food security and sustainability challenges. It is where digital technologies hold a universally innovative place to provide solutions to assurance, accessibility and availability.

The use of consumer goods is interwoven into all of the United Nations' 17 Sustainable Development Goals. While consumer goods improve the lives of billions of people, an inescapable outcome of their manufacture is increased consumption and utilization of natural resources. Consumer goods and sustainability have an uneasy relationship because their successful development increases demand and the requirement for the resources used to make them. Insights from the food-and-drink industry show how sustainability agendas have created breakthroughs in manufacturing and retailing practices that enable a rethinking of this view of consumption. For many consumers, it is perceived as a net depletion of resources, despite a legacy of sustainable reporting that demonstrates a circular economy is possible. This has changed and much of the catalysing of this has been due to the application of digitalization and automated capture of data across supply chains. Net-zero and carbon-neutral manufacturing is now possible, but reducing consumption when much of the world aspires to increase it remains antagonistic. The potential to embed sustainability into product development processes, so consumers act sustainably even when the supply is increased, is now identified in many supply chains. Food and beverage products have already demonstrated this with the application of indices for nutrition and food sustainability that have provided effective actions that improve food security. The methodologies to do this are applied to a national scale and there is a requirement to place them into all manufacturing operations.

We seek to demonstrate that the starting point for much of what we are trying to achieve is concerned with the consumer and an important part of this is describing what sustainable products and meals are. After working with food manufacturers and agricultural producers, across an accumulated six decades (the authors' careers), we know that any vibrant and sustainable business in agriculture and manufacturing must first recognize their customers' requirements. The thing that is most important above all is that food manufacturers – and the relationship between manufacturer and retailer – need to understand what we like to eat and drink more than any other operators in the food system. Understanding this relationship and utilizing the expert knowledge that goes with it takes us beyond the limits of traditional consumer or sensory science, and it is critical to future sustainability decisions. The food manufacturing and retailing experience is rarely spotlighted in a positive way in terms of innovation, sustainability or security; they are more often associated with impact or crisis. Processing of foods and presenting them to consumers is outshone by the dual importance of producing greater quantities of agri-commodities for a growing global population or defining the quality of diet needed for an increasingly unhealthy population. These two targets are typically driven by policy, not business, and while they do have the rightful causes of reducing hunger and improving diet, they often become woefully misguided and do not resonate with consumers. This is because the omission of the commercial functions of manufacturing foods and beverages and retailing clearly leaves inexcusable gaps in delivery of sustainability and security. This is recognized in the rise of the need to understand food systems as a holistic study and the fact that we have seen numerous reviews and investments made in delivering sustainable food supply. However, these have consistently fallen short of delivering across populations or the whole food system. We believe that if this continues, we will not meet the goals of increasing food sustainability and security globally. At the heart of all solutions is the processing and manufacturing functions of supply chains that create products and the meals we all experience daily. How we measure that product or meal as experts and consumers must have some form of common language if we are to improve consumption efficiencies. This must include all operators in the chain and it is a big idea and I am well aware that big ideas can be ignored because their solutions are often in contrast to an established dogma. The world of science is cluttered with big ideas that, once they were accepted into communities, became transformative because they were disruptive and possibly right all along. In our digital world, disruptive behaviours in commercial operations have always been present, but it is now the integration of thought leadership, increased transparency of idea-ownership (through social media) and commercial disruption driven by digitalization that provide the step changes that are put forward in this book.

This book raises many questions around our ability to assess and quantify the resources moving through the food system because new digital technologies offer many solutions to capturing and assessing data needed to do this. Connecting operations is crucial and providing a fingerprint of processes and products is now possible from farm to fork, or even concept to consumer with digitalization. A realization that this type of whole food chain assurance was possible started, for me, in 2008, when a team of researchers I was working with across UK and European groups were mapping the origin of biofuel feedstocks. Little did I know at the time that we were developing what we now call 'digital twins' and my specific tasks were to do this for the UK biofuel and wheat production system. There were concerns that biofuel production would limit food supplies and UK food manufacturers had innovative ideas for using more UK wheat and improving the quality of those wheats so that they could be used for baking bread. This requires high-protein hard wheats that are typically harvested in dry conditions, which are always unpredictable in the temperate North Atlantic corner of Europe. The Digital Twin demonstrator is shown in Fig. P1. It has projected the impact of biofuel production on bakery supply chains in the UK so that a robust and sustainable biofuel–food policy was developed for collaborative supply chains of food manufacturers, farmers and bioethanol producers (Martindale, 2009). The Digital Twin also provided evidence that recycling carbon through agricultural production systems and biofuel supply chains can reduce greenhouse gas emissions and improve air quality (by replacing fuel oxygenates in fossil fuels) (Martindale and Trewavas, 2008). The research and commentary around it was accepted in the most prestigious journals and this was a steep learning curve that I now look back on with an appreciation that my co-author stuck with me and my still relatively novice understanding of using data to define evidence. The initial research provided the means to develop spatial analysis of crop production and bioethanol refineries where land-use requirements for biorefineries were defined within 50 km radii of three national bioethanol refineries of the Vivergo, Ensus and Cargill companies. The Digital Twin made an estimate of the regional bakery demand for local grain based on local grain requirements of 36 bakeries, benchmarking their requirements of wheat with respect to annual financial revenue. The Digital Twin demonstrated that there were contingencies of over 0.5 million tonnes of wheat in this regional system that would support biofuel, feed and food production and this was used to guide sustainable food-system outcomes.

Many of the identified solutions to the biofuel and feed study required an understanding of how resources flow in supply chains with respect to specific time periods when most production of biomass and commodities are seasonal or transient. There was a great need for instantaneous data collection and interoperability between data sources, which was not thought possible in 2008 when I started this geographic information project. This instantaneous collection is now possible and holds much potential in auditing supply chain procedures in Industry 4.0 and 5.0, where operators now have an understanding of geographical information through the use of Google application programming interfaces (APIs) and so on. This was not the case in 2008 when geographical applications looked impressive but had little proven application in food systems. These datasets are often the limiting factors of the food system because knowing where things are in time and space make Fast Moving Consumer Goods (FMCGs) of value because they utilize resources to deliver specific convenience for limited time periods. That was where it started for me. The influence of geography and

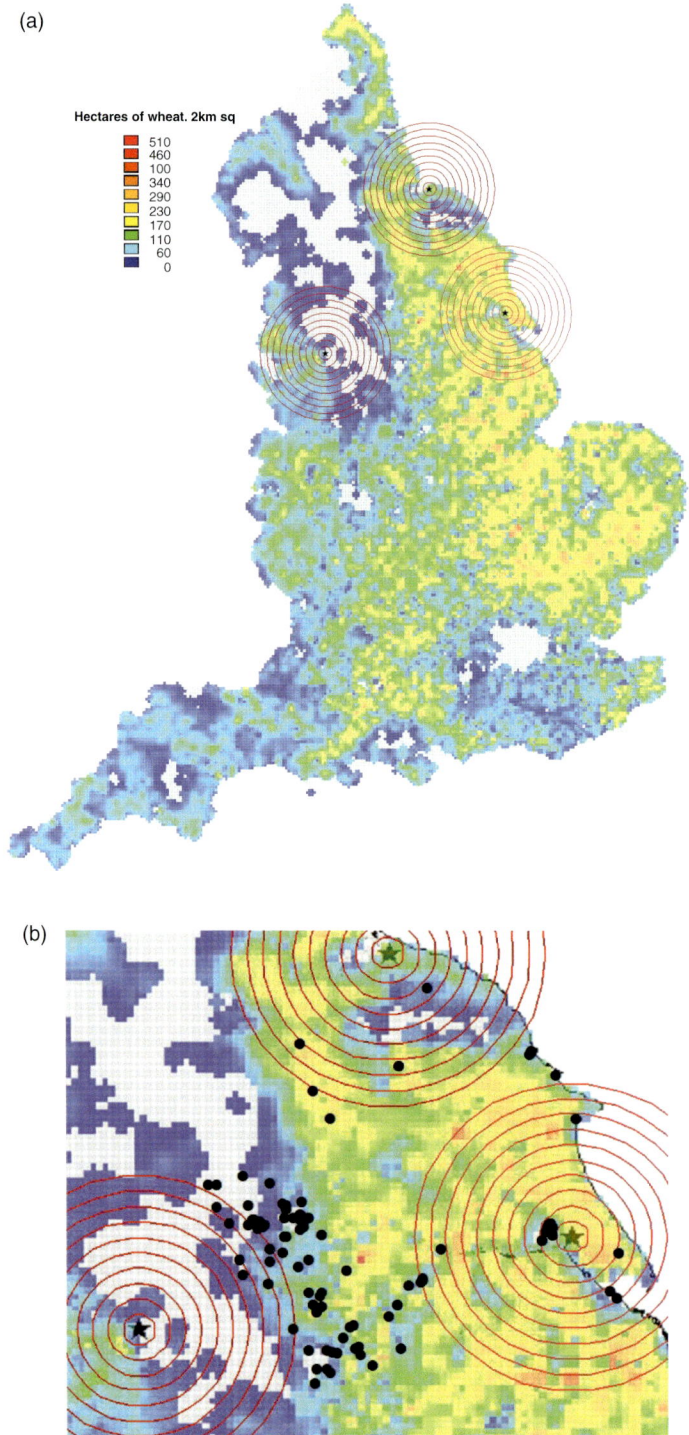

Fig. P1. An application of a GIS that extends to life-cycle assessment (LCA) and Digital Twin projections for biofuel resource flows. This demonstrator shows bioethanol production

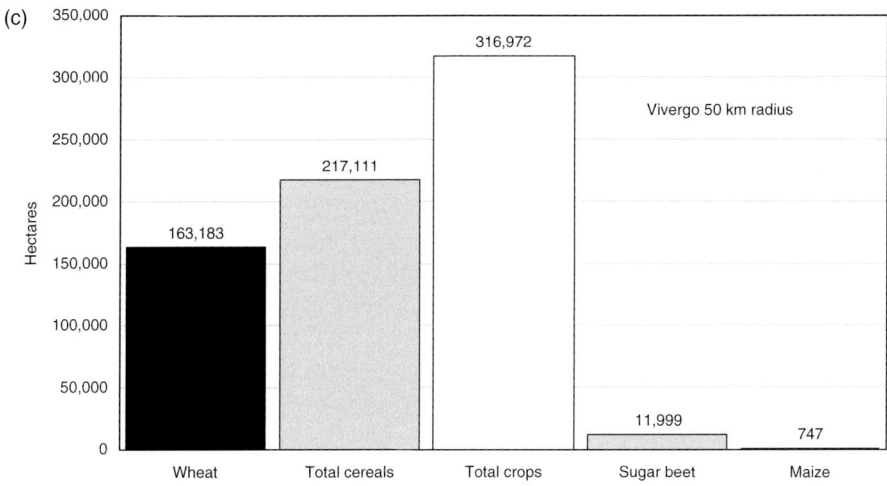

Fig. P1. Continued.
plants in England and their relationship to crop production within specific distances of the
biorefineries (a). The LCA outputs used in this demonstrator deliver projections for carbon
footprint, water use and any competition between food, feed and fuel supply. The concentric
circles are 5 km-wide circles with a radius of 50 km from the refineries. The grids show the
intensity of wheat production (red: 510 ha/2 km^2; blue: 0 ha/2 km^2) and this demonstrator
develops their relationship and connectivity with manufacturers. The black circles show the
location of major bakeries in these regions (b) (Martindale *et al.*, 2020) and crop biomass
produced within these regions suitable for fermentation (c). (Copyright W. Martindale, 2020;
developed in MapInfo Pro 10.0 and the grid square agricultural census data, as converted by
Edinburgh University Data Library, are derived from data obtained for recognized geographies
from the Department of Environment, Food and Rural Affairs (DEFRA), the Welsh Assembly
Government and the Scottish Government (formerly SEERAD), and are covered by Crown
Copyright.)

the large datasets associated with the movement of consumer goods is critical
because the food system has developed from global supply chains, whether the
food miles are low or not. This is because of our need to utilize semi-tropical
and tropical produce and because our chemical, mineral, land and labour resources are not uniformly distributed across the system.

The current globalized world-view was established some 35 years ago with
increasing economic freedom in eastern China. While these changes were disruptive in terms of economic and access to labour, the food system has always sourced from a global perspective because of consumer demand for food
and beverage products. An important aspect of understanding material flows
and utilization in the globalized world is that we build databases, and the
volume of data that we can access on material and resource utilization is increasing. Indeed, we have had Google for a generation now and the 'Google
generation' will have access to open data regarding lifestyles and culture in a
way that was not possible before 1998. The impacts of this are evident in the
way we access once-obscure or censored information to enrich our lives or

divert our attentions. There is clearly a 'good' and 'bad' side to this new world of open data. What is incredibly important to realize in the case of food supply is that connections between food and lifestyle are some of the most used drivers for obtaining data and information from external sources in this World Wide Web world-view. Our immediate task in food insights research is to provide a dynamic function to these maps, which are still static or snapshot in their current form – real time views and applications are yet to be fully integrated. Obtaining real time data has become possible through the use of blockchain and Internet of Things (IoT) applications, where applications regarding transport in cities and purchasing are already emerging.

Whether the information is used to guide diets or health is yet to be seen, but what has been very clear is that we consumers have access to dietary information. If we want to use a recipe we have never used before we can obtain it, free of charge, from the Web. However, in doing so we are also likely to be directed to a retailer database that will guide our choices for the most economical, most ethical and most quality-conscious ones so that we can make the perfect meal. It all sounds seamless and effortless – what could possibly go wrong with such a system? The integration of science and commercial and cultural aspects of food have become realistic goals in such a short time because IoT and digitalization have and will continue to transform our food system.

Wayne Martindale
FIFST, Editor in Chief
April 2022

References

Martindale, W. (2009) Co-development of bioethanol, feed and food supply chains that meet European agricultural sustainability criteria. *Aspects of Applied Biology* 95, 79–84.

Martindale, W. (2014) *Global Food Security and Supply*. Wiley, Oxford.

Martindale, W. and Trewavas, A. (2008) Fuelling the 9 billion. *Nature Biotechnology* 26, 1068–1070.

Martindale, W., Duong, L., Hollands, T.Æ. and Swainson, M. (2020) Testing the data platforms required for the 21st century food system using an industry ecosystem approach. *Science of the Total Environment* 724, 137871.

1 Our Connected Future and Global Food Markets

WAYNE MARTINDALE

1.1. A World Without Want[i]

A generation has passed since the publication of *Our Common Future*, also known as the Brundtland Report, in 1987. This report set out the need for indicators and quantifiable measures of the value of sustainable development (Brundtland *et al.*, 1987; World Commission for Environment and Development, 1987). At the time, the impact of globalization was unknown and this report bravely set out a call to action that provided Agenda 21, the rise of non-governmental organizations (NGOs) and, to some extent, the platform from which the current United Nations Sustainable Development Goals (SDGs) developed. It provided the global food system with targets or values for which baseline information was collected and future indicators and assessments of sustainable development could be made. This was important for food systems and the delivery of optimal nutrition for a global population that was projected to be 9 billion citizens within another generation. Indicators and assessment are a route to solutions, and they have provided a sense of vigilance for the global food system, but vigilance is not a call to action. We are now faced with the task of connecting our current understanding of the food system, where we know how much is produced, processed, consumed and wasted. Developing national indicators and assessment of international resource flows is well characterized, as are the methodologies that enable the footprinting of specific food categories and products. What is missing in all of this is the access to data regarding the sustainability value or external costs of resources by producers, manufacturers, retailers and consumers. If a consumer wishes to understand how a product has been selected from farm to fork, the data concerning greenhouse gas (GHG) emissions, resource losses and water use are typically available but unseen. There are several reasons for this, such as confidentiality of proprietary data, regulatory compliance and competition regulations. It is from within

© CAB International 2022. *Food Industry 4.0: Unlocking Advancement Opportunities in the Food Manufacturing Sector* (W. Martindale *et al.*)
DOI: 10.1079/9781789248593.0001

these commercial frameworks that the need for sustainable reporting is proving disruptive because there is a need for transparency in claims and information associated with food and beverage production, manufacture and retailing.

The ability to connect different food supply chain data sources from farming, processing and manufacturing operations (not typically visible to consumers) to the distribution and retailing operations (that are visible to consumers) has become possible with enterprise resource platforms (ERPs) that manage resource inventory and control orders across a business. These are all confidential and remain within sight of the business, but the emergence of certification systems has changed who has the need or right to have access to these data. Examples of certification systems are shown in Table 1.1. They have transformed transparency and reporting of practices – of this there is no doubt – but a question for our new digitized food system is: will this need for transparency further disrupt practice? Much of the pressure to embed transparency has been in response to global food safety improvements and consumer concerns regarding just or fair products. The emergence of online media and the use of social media by consumers has raised the profile of the source and quality of foods, creating movements that have specific values associated with them. Examples of the initial movements include those that wish to exclude industrial farm inputs, such as those for organic or biodynamic foods. While these movements emerged at the beginning of the 20th century in response to industrialized agriculture, social media has changed their impact. The cause of these movements has extended to the exclusion of biotechnologies and genetic modification. This exposed much of where our current need for transparency came from because there was a requirement for companies to disclose information that previously could have been kept confidential and even undisclosed to audit of any kind. Media, and in particular social media, have had an important role beyond transparency to that of trust and complete disclosure of any information.

1.2. The Need for Transparency in Our Global Food System and the Opportunity of Digitization

The need for disclosure is becoming less of a barrier to such whole supply chain transparency because of digitization and the use of blockchain systems. Simply put, blockchains can detect when data or information is not reported or it is incorrect. There is nothing new with this – forensic accounting has done this for many years and it is often said that irregularities in supply chains are often first detected through financial faults. What blockchains do very differently is make sure the data placed into the blockchain system is sharable, interoperable and remains in this form forever; it is immutable. This provides a mechanism for trust to embed. The reporting of resource flows in food supply chains has highlighted the importance of data flows to assess how connected operators in supply chains are. Resource inventories are the glue that can hold the whole food system together and data resources are critical in determining whether information is correct and can be trusted. Systems such as blockchains

Table 1.1. A comparison of selected internationally recognized certification schemes.

Certification scheme	Organic	Fairtrade	Rainforest Alliance	UTZ Certified	MSC	Proterra
Regulated	In 49 Nations	No	No	No	No	No
Accredited certification	ISO65/BS EN45011		ISO65	ISO65/BS EN45011	ISO65/BS EN45011	ISO65/BS EN45011
Accredited standards	ISEAL compliant	ISEAL compliant	ISEAL compliant	ISEAL compliant	ISEAL compliant	
Whole chain certification	Yes	Yes	Chain of custody	Chain of custody	Fishery certification and chain of custody	Yes
Environmental standards	Yes	Yes	Yes	Yes	Yes	Yes
Animal welfare standards	Yes	No	Cattle only	No	No	No
Ethical/social standards	In IFOAM principles	Yes	Yes	Yes	Yes	Yes
Synthetic agro-chemicals	Highly restrictive	Minimal and responsible use	SAN list of prohibited pesticides	Responsible use	Not applicable	Requires minimal use
Genetic modification	Prohibited	Prohibited	Prohibited	Not stated	Not applicable	Prohibited
Global sales value	€559 billion	€4.9 billion	Not stated	Not stated	€3.5 billion	Not stated
Global area of production	37 million ha	Not stated	75 million ha forest and 1.53 million ha farmland	Not stated	8% of wild-caught fisheries	Not stated

Notes:
ISO65 Agriculture, see https://www.iso.org/ics/65/x/
EN45011, BS EN 45011:1998 General requirements for bodies operating product certification systems
ISEAL, see https://www.isealalliance.org/
IFOAM, see https://www.ifoam.bio/
SAN Sustainable Agriculture Network http://san.ag/web/

and distributed ledger technologies (DLTs) can distribute responsibility for trust (Martindale *et al.*, 2018a). This can be demonstrated for carbon footprint data that can provide a measure of supply chain efficiency for processing and manufacturing inputs from farm to fork or farm to taste. The carbon footprint is an appropriate means to report energy use and resource flows for many food and beverage products and it can be used to communicate impact across supply chain functions, including consumption. This is because there are specific data inputs for each activity associated with a footprint. As an example, a typical 200 g mixed livestock and plant ingredient sandwich will have 220–290 g of GHG emissions associated with growing and processing its ingredients, transport and packaging will contribute 20–50 g GHG emissions, and GHG emissions such as methane (from livestock production) and nitrous oxide (from organic and mineral nitrogenous fertilizer use) can significantly increase these emissions. The GHG emissions can be reduced by fit-for-purpose agronomic management and efficient supply chain planning, which depend on efficient data management (Martindale *et al.*, 2018b). If such data placed in a blockchain have been geocoded with a location and time reference, this provides further proofs for the farm-to-fork view of supply and provides an integrated Geographic Information System–life cycle assessment (GIS–LCA) method (Martindale, 2017). The GHG emissions associated with foods are becoming critically important to report and have been identified by the USA Environment Protection Agency as potential targets for full assessment as Scope 3 GHG emissions. These are those associated with the value chain. Digital technologies have demonstrated how DLTs and blockchains can track high-volume and high-variability financial flows as immutable data. It was once thought Scope 3 emissions were far too variable and complex to deal with, but the blockchain approach now provides a solution. This would provide an instantaneous assessment of resource flows in food and beverage supply chains. These types of value chain processes have a role to play in how much product is likely to be used and wasted, so their impact is significant (Martindale *et al.*, 2020a).

The approach can be packaged as Digital Twin systems (DTSs) for food supply chains that provide instantaneous and incisive analysis of resource flows. But they are still dependent on quality data, otherwise there is the possibility of garbage in–garbage out (GIGO) scenarios. Digital solutions such as blockchains are making this less likely, but it is still a risk. The potential to falsify or invent data will always exist and there is a requirement to build a culture of trust for any data input. Blockchains do this by making data immutable, so that good data (trusted) and bad data (potential lies) are always on the blockchain (Rejeb *et al.*, 2020). Technological fixes such as blockchains can also make data input accessible, so they complement structures of trust that will ultimately be developed through the establishment of communities, collectives and business ecosystems. LCA and carbon footprinting are routinely used to assess the environmental impacts and wider sustainability reporting of products where there is an increasing need for trust, because LCAs are used to report claims such as those associated with carbon-neutral products (Martindale *et al.*, 2019). The use of traceability or transparency software solutions such as blockchains that can trace the LCA or footprint data from source to product to consumer do help overcome these limits and gaps. In many respects the blockchain approach and the footprinting approach are similar in that the flow of trust and materials must balance – that is, what data or material goes into a system must come out, it must be 100% in and 100% out in terms of mass-flow. If the 100% balance is not achieved, the blockchain will flag where imbalances in material flows or trust are in the system; the difference between this and an LCA is they are immutable and in the blockchain forever.

Global events demonstrate time and time again that supply chains are resilient, with food supply and consumer demand being finely tuned, but not as flexible or agile as many consumers in Europe have come to expect because of the impact of untrusted data on commercial claims. The indicators of global food supply and price show primary food commodities, including small grains, oils and dairy products, are volatile, but it is trusted real-time data that are often the most limiting in terms of the speed that this system responds. This is despite the impact of digitization and the new applications of blockchains, which is counterintuitive because technology should improve outcomes, but the major limitation remains collaboration between companies across supply chains. An example is provided by production and trade data, which are typically reported annually, so forecasting in business is typically made using annual assessments of the financial worth and mass volume of resource flow. Much of the data required to report this on a day-by-day basis are owned by suppliers themselves. So it is exciting and innovative to consider the very blockchain tools we are beginning to use could start to enable collaboration and benchmarking across supply chains so that they are agile in responding to day-on-day changes that are the reality of modern trading and potential crisis. While many of the currently used blockchains have developed to enhance assurance and reduce the risk of safety failures, the reach of their applications goes much further in that they are secure and gated collective sources of supply chain data.

1.3. The Requirement for Food Baselines and Prior Art

The application of global indices of nutrition and food sustainability in public health and the improvement of product profiles have facilitated effective actions that increase food security. We develop index measurements further here so that they can be applied to food categories and be used by food processors and manufacturers for specific food supply chains. The research considers how they can be used to assess the sustainability of supply chain operations by stimulating more incisive food loss and waste-reduction planning. It demonstrates how an index-driven approach focused on improving both nutritional delivery and reducing food waste will result in improved food security and sustainability. Nutritional improvements are focused on protein supply and reduction of food waste on supply chain losses and the methods are tested here using the food systems of Kenya and India. Innovative practices will emerge when nutritional improvement and waste-reduction actions demonstrate market success, and the co-development of food manufacturing infrastructure and innovation programmes will result from them. The use of established indices of sustainability and security enable comparisons that encourage knowledge transfer and the establishment of cross-functional indices that quantify national food nutrition, security and sustainability. The research presented in this initial study is focused on applying these indices to specific food supply chains for food processors and manufacturers and using them to provide further insight.

Current indices of food security and sustainability are designed to identify high-level policy risks at national scale and as such they focus on the production of agricultural commodities, providing little understanding of resilience in the manufacturing, distribution and retailing functions of supply chains. This is important because it is the development of industrial infrastructure that has consistently delivered resilience in food supply chains when projections of 'peak resource' models operating at national scale have not delivered accurate projections. This is because the models based on limiting carrying capacity overlook values associated with innovation in supply chains, such as those of contractual trust and organizational cultures that deliver the commercial goals of supply. It is increasingly important to understand resource limits of food production systems with respect to these commercial attributes and improving the tools to do this remains crucial. The measurement of the associated innovation values in small and large manufacturing companies alike are focused on providing the consumer fulfilment of manufactured foods. Their value is rarely mentioned as a contribution to sustainable food supply and the need for manufacturer-relevant assessments has to be established for greater food security. Food security itself is defined as the state in which people at all times have physical, social and economic access to sufficient and nutritious food that meets their dietary needs for a healthy and active life. This framework is based on the internationally accepted definition established at the 1996 World Food Summit. The interaction of food security with nutritional goals and food supply-chain sustainability do complement each other and focus the requirement for food industry guidance at product development level. The global food system in 2050 will need to supply 9 billion people with meals in a safe and sustainable way that provides all the attributes of a secure food industry.

1.4. Eco-design and Co-creation

As well as meeting sustainability and nutritional goals, food products must be affordable, available and assured so that they can be prepared for meals, which raises the importance of understanding the role of new product development (NPD) in the global food system (Martindale, 2017). It is here that NPD processes must take a concept-to-consumer approach, so that sustainability is built into the NPD process and food waste is removed from the food system. The concept-to-consumer design approach will mean food products will be developed for utilization in meals and therefore result in improved nutritional, food waste and sustainability outcomes. Product developers are keystone operators for enabling food sustainability and they must begin to take a long-term view for continuous improvement in these processes. This will require a step back from typical NPD operations to take stock of what successful product development means for consumers and their diet at a population or meta-NPD level (Martindale *et al.*, 2019). Understanding resource flows in the food system and into NPD processes is important because any co-creation for products will be limited by resource accessibility and affordability. Understanding where critically traded resource flows are in a food system is an important aspect of any NPD strategy. In the case of food and beverage products, it is especially important in temperate regions that depend on semi-tropical products for fresh produce that can either not be grown in temperate areas or provide out-of-season supply. Semi-tropical regions also supply luxury ingredients such as coffee, tea, chocolate and spices, which are essential to any NPD strategy. Proof of this principle is provided for the criticality of the supply of flavourings such as vanilla, stimulants such as tea and coffee and semi-tropical fresh produce.

Figure 1.1 shows the most valued agri-food imports into the UK from India and Kenya. The greatest value is associated with those products that will have greatest demand as ingredients or products. Figure 1.1 demonstrates the trade of high volumes often follows a Pareto relationship, where much of the volume or value is associated with relatively few product categories. The connectivity between trading partners in the food system will need to be considered if sustainability and security goals are to be met and this requires greater incisiveness in the use of geo-spatial data (see Chapter 2). The data for India and Kenya demonstrate rice, tea, green beans and plant materials (e.g. plant stocks and bulbs) are the most important categories imported into the UK with respect to financial value. These relationships suggest that these are the categories of priority with respect to food defence, but such an approach undermines any food defence policies because they should act across the food system. The highest value categories are where there are increased risks, but lower value categories represent the same risk to food defence systems because breaches in them are carried out by opportune activities that can often be considered rogue elements. These are often associated with other activities such as meeting contractual conditions, limitation of resources and increases in demand – all of these can occur within the law of our global food system. Understanding the availability and demand for these categories of ingredients is an important starting point for any NPD strategy that will identify sources of ingredients for development. In doing so, the approach will build in sustainability, nutritional and food defence considerations, making the outcomes more likely to meet customer and consumer demands. Developing the strategies that identify criticality in the food system provides an important platform for evidence on sourcing that can support analytical processes and blockchain systems.

Fraudulent activity in food and beverage supply chains is of considerable importance, not only in tackling social compliance issues such as modern-day slavery or sustainability reporting, but also with regard to direct counterfeiting or contamination of products for financial gain. It is these issues that are of primary importance in tracking and traceability of food materials and it is time to consider how sustainability metrics can be aligned to them (see Chapter 2).

(a)

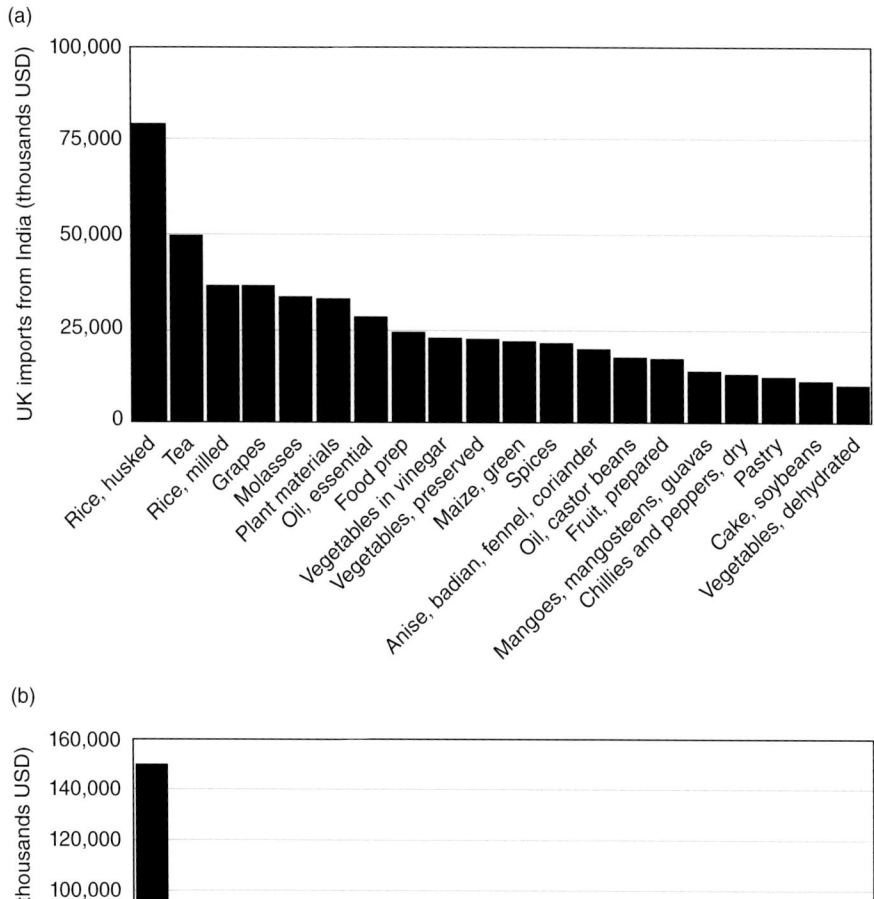

(b)

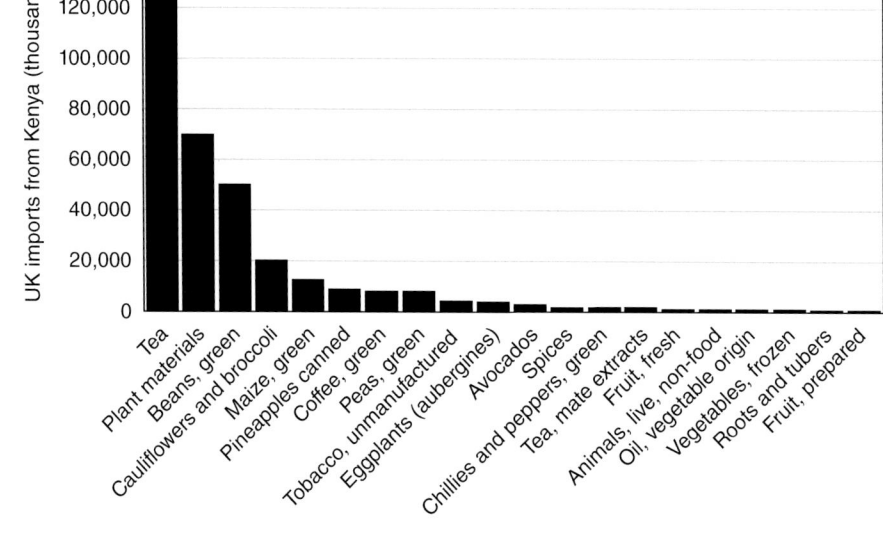

Fig. 1.1. The imports of the most valued 20 crop and livestock products imported by the UK from (a) India and (b) Kenya in 2019, reported in thousands of US$. (Data derived from FAOStat trade and commodities balance data)

Table 1.2. The selected food defence risks associated with beverage supplies where high value in supply chains exposes risk to fraudulent activity. It should be noted that false geographical origin declaration is an important source of fraud, where the potential to target high-quality wine brands is one of the most important target categories.

Beverages	Food defence risk
Alcoholic beverages	Due to their high market value, many brands of wine and spirits are at risk of being counterfeited, mislabelled and adulterated. Dilution with water or cheap alcohol is a common fraud, as well as addition of undeclared sugars or sweeteners and even substitution of the beverage with a lower value one.
Cocoa and chocolate	Chocolate compounds are exposed to the risk of being adulterated with exogenous materials, or mixed with fats of different species origin to improve the flavour. In some cases, producers or distributors were even found to be adding artificial or different origin material to cocoa powder in order to increase its weight.
Coffee	Coffee powder and soluble coffee are likely to be mixed with exogenous materials to increase the content weight or adulterated using undeclared sugars and sweeteners. Cases of mislabelling are also quite common, especially false species origin declaration (*Robusta* variety declared as *Arabica*).

Table 1.3. The selected food defence risks associated with ingredient supplies where high value in supply chains exposes risk to fraudulent activity. The designation of ingredients here is important in that they were selected because they occur in many manufactured foods as an ingredient or component, which compounds the issue of providing traceability. The use of geographic origin remains crucial and European high-value dairy products (e.g. Parmigiano Reggiano cheese) are defined by PDO protocol requirements that can be violated.

Ingredients	Food defence risk
Eggs	Many cases of food fraud concerning the production of eggs involve the addition of exogenous – and sometimes harmful – substances (e.g. artificial dyes, oils or fats), aimed to improve the product's flavour, appearance, etc. False declarations of organic farming or geographical origin are a common risk, too, as well as the fraudulent substitution of fresh eggs with undeclared incubated or frozen-thawed ones.
Honey	High-value honey varieties, such as Manuka honey produced in Australia and New Zealand, are exposed to the risk of being adulterated by exogenous sweeteners to improve their flavour or substituted with low-quality material. Cases of mislabelling have also occurred (false declaration of geographical, botanical or animal origin).
Milk	Milk formulations such as milk powder are prone to be diluted or adulterated with exogenous fats or oils; false geographical origin declaration is another common fraud.
Oils	High-value oils such as extra-Virgin olive oil are often subject to addition of different oil varieties or even complete substitution. In some cases, artificial compounds are also employed to adulterate oils.
Spices	Adulteration by undeclared artificial dyes or flavourings is one of the most common frauds in dried and powdered spices; substitution with other plants is high risk.
Vinegar	High-quality brands of vinegar (Balsamic vinegar of Modena, for instance) are subject to the risk of dilution with water, or complete or partial substitution with lower value material.

Table 1.4. The selected food defence risks associated with food products supplies where high value can expose risk to fraudulent activity. The designation of products here is important in that they were selected because they are processed, packed or manufactured foods and can deliver whole product claims. It should be noted that false certification declaration is an important source of fraud.

Products	Food defence risk
Fish	Mislabelling (e.g. false species/geographical origin declaration, farmed fish declared as wild) is among the most common frauds. Substituting high-quality species with lower value or even potentially toxic ones is also a widespread practice, and so is temperature abuse. In some cases, traders and distributors have been found breaking the regulations concerning storage temperatures, or declaring frozen-thawed fish as fresh.
Cereals and flour	The high-risk frauds include addition of exogenous compounds (e.g. melamine, used to apparently increase the product's protein content), false organic farming declaration and partial or complete substitution with different species. A typical case of counterfeiting consists in the substitution of durum wheat (*Triticum durum* – the most suitable cereal for the production of pasta) with lower value soft wheat.
Fruit, derivatives	Many cases of adulteration, mislabelling (e.g. false declaration of geographical origin or organic farming) and dilution concern fruit-based preparations and especially juices; high-price juices are most at risk of being diluted or partially substituted by material of different species origin. Undeclared exogenous sweeteners might be added to fruit compounds to enhance the flavour.
Meat	Among meat-based preparations, meatballs and burger formulations are the most vulnerable to adulteration, mislabelling and partial or complete substitution with different, lower value or potentially toxic species. False geographical origin and false organic breeding declaration are high risk.
Mushrooms	Among the high-risk frauds concerning mushrooms and mushroom-based preparations are the addition of exogenous material in order to enhance the product's appearance, adulteration and substitution with lower quality varieties or species and, naturally, mislabelling frauds such as false species origin or organic farming declarations.
Tomato, derivatives	Tomato juice and other similar compounds are particularly vulnerable to adulteration or substitution with overripe or GMO material. Mislabelling frauds such as false declaration of organic farming or species origin are high risk in food defence.

Examples of these assurance driven activities are reported in Tables 1.2–1.4, for beverages, ingredients and foods, respectively. The examples demonstrate the need for data trust in the food system because food and beverage supply chains must be defended from the action of what are rogue elements or mistakes made that will often cross legislative boundaries and actually contravene the law. The data trust frameworks that blockchain and other systems can embed need to be supported by programmes that provide analytical proof of origin and composition by effectively fingerprinting a product to a specific source and demonstrating product integrity. Such fingerprints can be strengthened by on-pack labelling and assessment of in-pack product composition or quality. The data being presented here have the potential to be transformative because of the need to meet the UN SDGs, carbon-neutral targets and environmental standards across the food system. Product development will increasingly be asked to link the constraints of food product design with high-level targets, such as the UN's 17 SDGs that cross the health, social wealth and environmental lenses of sustainable food supply (Casini *et al.*, 2019). Consumer choice should be at the core of getting food product development right. A starting point for developing NPD strategies guided by these principles is to use national-level consumption statistics to provide guidance on the most popular food choices. Switching or nudging consumer choices of foods to more sustainable dietary options is not only possible, it has already happened for protein choices, where it was considered unthinkable not that long ago, and has been driven by consumer demand (Sachs, 2012). Many NPD functions are trying to catch up with this shift in consumer attitudes and the diversification of protein choices is an important focus for future dietary sustainability in the production and export arenas.

1.5. Supply, Demand and Ecosystem Services: The Death of 20th-century Food Economics

An analysis of the demand for different protein categories means associative data analysis techniques are required and they have been tested to identify pressure points associated with the movement of resources within and across supply chains. It is also the nutritional quality of food materials that determines demand, and these are rarely considered in trade data because the importance of volume and price alone dominates reporting. Protein content is used as an indicator of nutritional value of crops, such as pulses and grains, and it is used in the Centreplate Model developed by research that can rank food materials based on their protein supply in diets (Martindale *et al.*, 2020c). It is the measurement of nutritional value and the risk of food being wasted that provide important outcomes for consumption because if food is not fully utilized any resources used to manufacture it are lost. It is this insight on the utilization of foods by consumers that is the core principle here and it conveniently connects the sustainability attributes of nutritional improvement and waste reduction, which are universally desirable impacts across supply chains (Martindale, 2016). In this study the association with protein supply, as a nutritional benchmark, is tested. The amount of food waste associated with different food categories is obtained from FAOSTAT data.

There is currently a recognition that improvements in obtaining supply chain data will create a step change and digital technologies offer much promise in improving data capture by operators across supply chains. That is, local data and bespoke data captured by food companies could be of great value in future and the use of indices of security and sustainability would result in reportable good practice (Martindale *et al.*, 2018a). Real-time supply chain data is a future capability that can be provided by recently tested applications of cloud-based data transfer that is secured by DLTs in the food industry. When the algorithms for assessing sustainability become transferable to all supply chain partners, it is the communication of them that becomes guiding. This is dependent on scaling data to the national and global food marketplaces, where the ability to obtain verified and transparent data for products in these supply chains has previously limited sustainability assessments of consumer goods. This is now being overcome using technologies that tag products using data carriers such as optical characters and bar codes that ensure secure supply chain records are available. The application in sustainability is only beginning to be tested here. In the case of waste reduction, more efficient inventory planning will mean quality is maximized and dramatic reductions in household food waste are observed. Of course, the data for consumer utilization of food will still limit data in the supply chain and the use of digital applications associated with purchasing are already making this consumer feedback possible.

The geographical and time origin of a product are important aspects of traceability. Figure 1.2 demonstrates how relatively simple source data can rapidly become complex and chaotic. The data used for this mapping come from the FAOStat databases on detailed trade. Figure 1.2 shows the amount of avocados imported into the UK and where they are produced, from 13,000 tonnes up to 26,000 tonnes for the largest source or trade partner countries such as Peru, Spain, Mexico, South Africa and Israel. There are also partner countries that are the source of up 2600 tonnes of avocados imported into the UK each year. Figure 1.2 shows the Pareto scenario again here in that the high-volume importers will be associated with high sensitivity to food defence issues. This has occurred with avocados, where in Mexico their production has been associated with illegal labour practices and deforestation of high conservation areas such as virgin rainforest. However, lower volume producers must not go unscrutinized, even though the resource flow is ten-fold less, because this is where compliance can go unchecked. Figure 1.2 also shows how importers are not necessarily producers, with Germany importing up to 13,000 tonnes of avocados into the UK in 2019. The role of secondary importing nations is crucial for air freight, deep water ports and primary processing (e.g. ripening of fruits). Certification in such a diverse range of import volumes is essential and making sure that every producer is complying with certifications brings many other issues. The evidence for compliance requires auditing and it means that an individual must currently witness practices so that they can be verified as complying. Digitization and blockchain systems offer an opportunity to begin to consider remote audit options so that instantaneous audit of supply chains becomes possible if trust is assured. The food industry is responding to this. It is recognized that an integration of digital, chemical analysis and existing auditing practices that verify practices can bring us closer to more accessible and affordable methods.

The example presented in Fig. 1.2 shows the largest resource flows (greatest amount of imports) are from relatively few countries but several low-volume importers also provide the same food defence risk, and this exists for other food categories. It presents a universal challenge in food systems, whether food

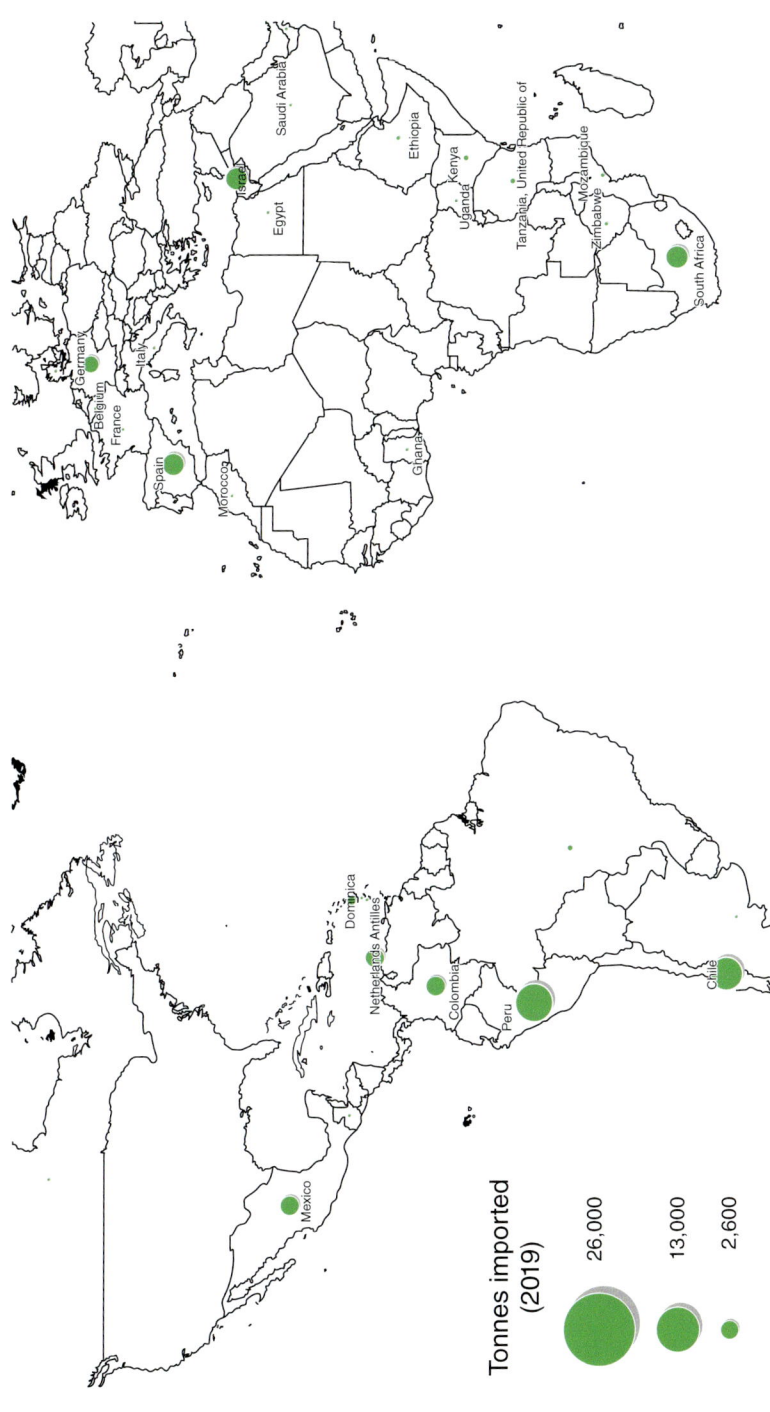

Fig. 1.2. The import of avocados into the UK (tonnes) in 2019. (Data derived from the FAOStat Detailed Trade Matrix database; Base World Map image is the intellectual property of Esri and is used herein under licence. Copyright © 2020 Esri and its licensors. All rights reserved.)

defence, compliance or sustainability is assessed, and it is even more apparent where there are several processing operations, such as from cocoa to chocolate. Figure 1.3 shows the import of cocoa beans into the UK is from relatively few producers and secondary importers that do not produce cocoa beans (Fig. 1.3 A). However, cocoa beans are processed into butter, powder, paste and cake that is imported from different nations (Fig. 1.3 B). The processing of cocoa beans is where financial value is increased, and the volume of cocoa butter imported compared to cocoa beans halves, but the value of butter is increased. A similar relationship exists for pastes, cakes and powders, where value is increased through processing into ingredients.

A similar situation exists for many manufactured food categories in which processing operations change how resources are traded. This is notable for meat products, where there have been breaches of food defence. The import of beef into the UK is dominated by imports of whole carcasses and finished cuts of close to 200,000 tonnes in 2019 from Ireland, for example, but some 20,000 tonnes of prepared beef products are also imported from Brazil (e.g. beefburgers). Most prepared beef imports are from Ireland and Brazil into the UK. This demonstrates it is crucial to identify where food defence risks are in a food system and the risks associated with imported prepared beef into the UK from Brazil are typically visible when issues regarding deforestation are raised. Transparency across these higher risk supply chains and imports will require enforceable digital assessment of geographic origin and their compliance with international certifications.

1.6. Methods for Assessing Food Utilization that Require Internet of Things Interventions

Understanding how resources move in the global food system at national scales is crucial if assessment of national efficiency and utilization of food resources needs to be obtained. The requirement for this is evident if the sustainability of national food production from farm to fork is investigated. If it is not done, studies do not work to a baseline measurement and they are unable to benchmark to international assessments such auditing or certification. The alignment of practice to the UN SDGs is critically important and without knowing where to start to measure efficiency and then to assess what data means, this is a futile exercise. For example, the main three proteins supplied to the UK population are wheat, dairy and poultry. It is sensible to use supply as an indicator of those food categories that are of most importance to the UK system. This principle is developed in the Centreplate Model reported previously, which benchmarks priorities for assessing food supply and wastes based on protein value and domestic supply quantity (DSQ; production and imports minus exports and stock variance) of specific food categories (Martindale *et al.*, 2020c). The metrics and algorithms used to associate different food categories are tested in this research and they can provide communications that resonate with consumers. All external resources used to manufacture and distribute food are focused on the

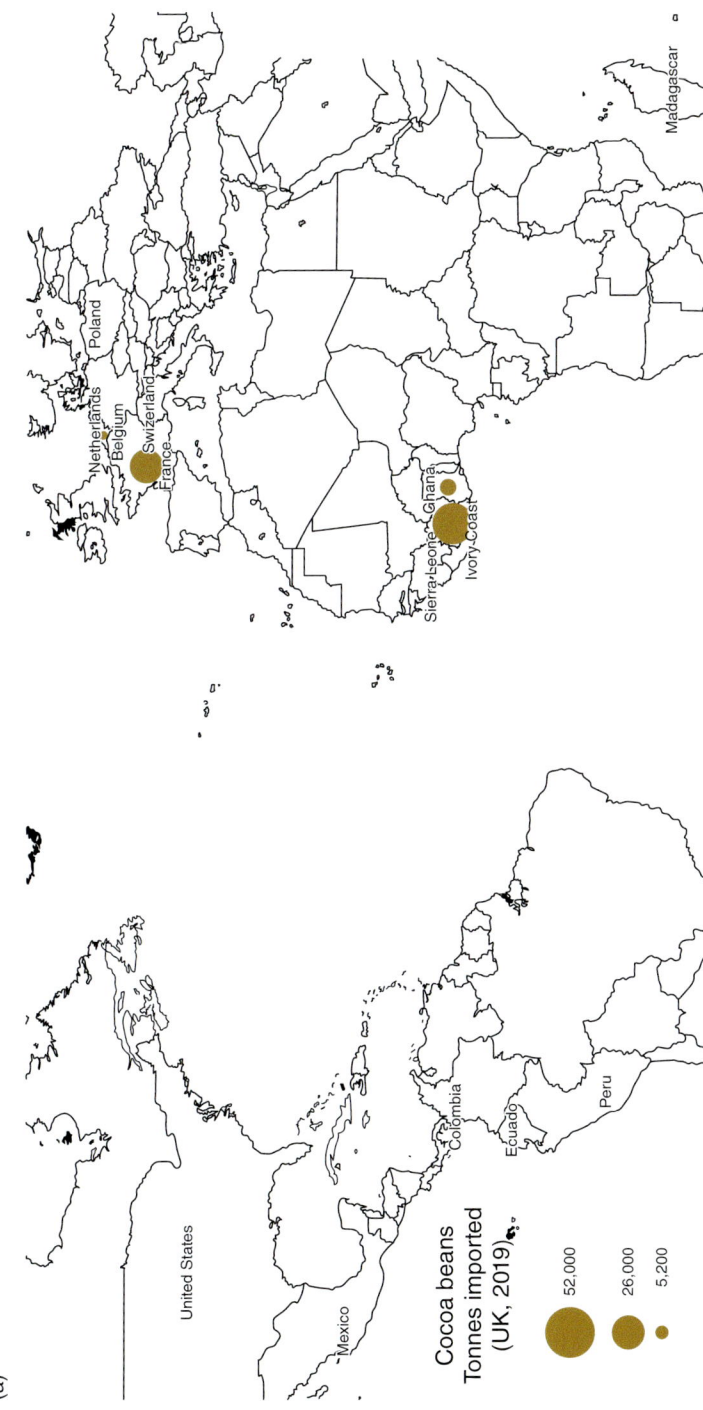

Fig. 1.3. The import of (a) cocoa beans and (b) cocoa butter into the UK (tonnes) in 2019. (Data derived from the FAOStat Detailed Trade Matrix database; Base World Map image is the intellectual property of Esri and is used herein under licence. Copyright © 2020 Esri and its licensors. All rights reserved.)

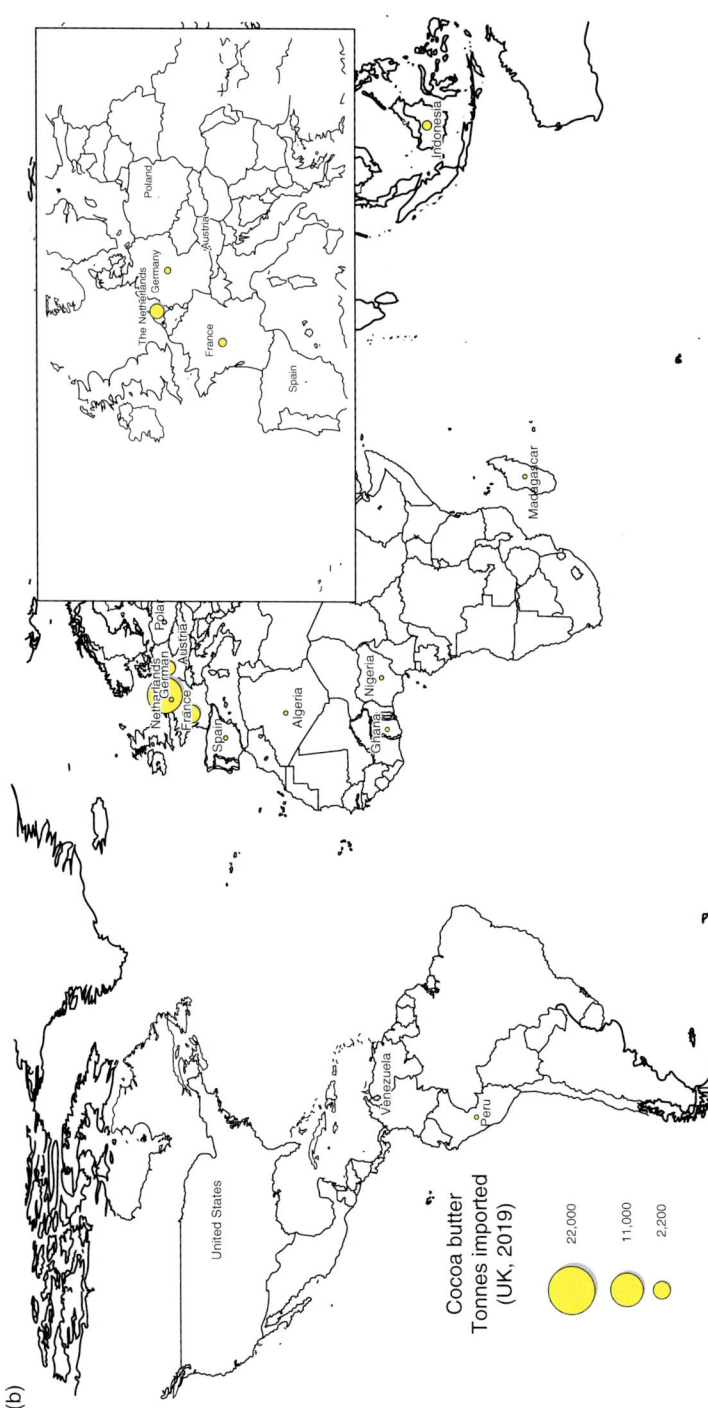

Fig. 1.3. Continued.

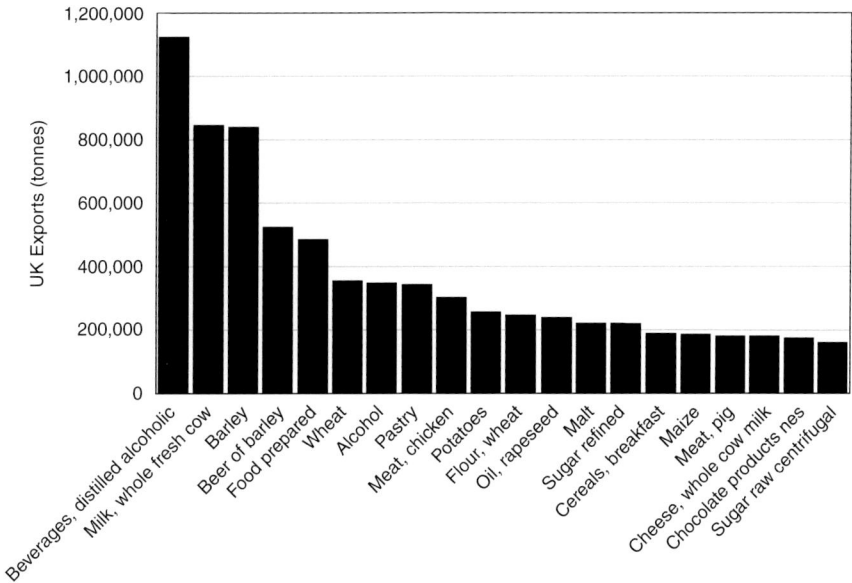

Fig. 1.4. The exports of the most valued 20 crop and livestock products exported by the UK in 2019, reported in tonnes. (Data derived from FAOStat trade and commodities balance data.)

final experience of consuming food and beverages. The innovative approach to averting a resource use crisis based on this research is to make sure that product fulfilment meets the sustainability requirement of food products when they are developed using the methods reported here. The relationships established for imports also stand for exports and this is seen in Fig. 1.4 for exports from the UK.

Established indices include the Global Food Security Index (GFSI), Global Access to Nutrition Index (ATNI) and Food Sustainability Index (FSI; Gustafson *et al.*, 2016; Chaudhary *et al.*, 2018; Haddad, 2018; Chen *et al.*, 2019). These have provided platform studies for sustainability indices so that the impact of these high-level indices can be extended to individual manufacturers and product developers. In the UK there are over 350,000 food businesses. The goal of a Sustainability Index is for it to be applied to improve security and sustainability in each of these food businesses when products are at concept, development or manufacturing stages. This would result in the delivery of sustainable outcomes that are transparent and interoperable across supply chains, which is something that existing LCA approaches do not provide. The application at such an operational level must be open sourced and available to food businesses because there are known barriers where current indices for nutrition, security and sustainability exist, but they are only typically used by large food groups and companies. In an open data food system, access to methods for assessing the sustainability of food products must become available across the food system and this is often limited by resourcing and skills. Certifications offer an example where application can be limited, because even though they have transformed practices the potential of 100%

certification has only been shown for foods such as bananas (Wilson and Jackson, 2016). In others, such as chocolate and coffee, an increased market share is seen compared to non-certificated products. However, they still only account for less than a fifth of total food production in other markets (Grunert *et al.*, 2014). In the cases of the Marine Stewardship Council (MSC), 12% of wild fish caught with even less ocean area under protection (Lester *et al.*, 2013; Sala *et al.*, 2021); Roundtable on Sustainable Palm Oil (RSPO), 19% of palm oil suppliers; Roundtable on Sustainable Soybean (RTSS), 1% of the global production is certificated (Rueda *et al.*, 2017). The goal sounds simple. It is possible to embed responsible values in a process from start (producer) to finish (consumer) in a specified supply chain and scale this activity so that there is a 100% responsible, ethical and carbon-neutral shopping basket for a sustainable diet.

The approach of relating nationally reported metrics to diets has been tested at a food category scale using the Centreplate Model using national food balances that operate at the scale of diets rather than national output and seeks to consider the realistic protein diversification of diets (Martindale *et al.*, 2020b). It has exposed the importance of such tools in the NPD functions of food manufacturing companies because new protein ingredient categories are required to be available at national scale. Applying the model has highlighted the ability to assess the value of NPD with respect to nutrition, distribution and food loss, but the application to product development was identified as an area that could be further developed for sustainability outcomes. A transformative approach here was to consider how food and beverage products are utilized by consumers and use this to feed back data associated with utilization into the product development processes so that product design essentially builds-in sustainability and weeds-out food waste. The measurement of meals being wasted provides important outcomes for consumption, because if food is not fully utilized any resources used to manufacture it are lost. It is this insight on the utilization of foods by consumers that is the core principle here and it conveniently connects the sustainability attributes of nutritional improvement and waste reduction which are universally desirable impacts across supply chains (Martindale, 2016). The development of the Centreplate Model has identified the requirement for a robust and accessible analysis of sustainable product development that can be used by food manufacturers and processors. This approach was developed as a model that used six attributes or functions that defined protein content of foods, distribution of products, energy embodied in their production and processing, waste associated with their use and GHG emissions (Martindale, 2017). What was specifically important in these models was the use of meal concepts developed by chefs. The method of categorization of meal types was derived from chefs and cooking books (one in particular by Jamie Oliver, *Jamie's Ministry of Food* (2008), which arranges meal types in a way that is robust and accessible). The route to meal and categorization for nutrition and sustainability has been explored by others, including the Australian Total Well Being Diet that demonstrated it could be achieved and placed nutritional sciences into meal planning (Noakes and Clifton, 2005). The study has shown possible applications and routes to simplification where the GHG emissions of products are related to their nutrient density, risk of food waste and distribution. The need for robust and accessible analytical tools in the manufacturing sector is important because if sustainability is built into NPD, it will need guidance for every food and beverage product.

The GFSI has been used to provide the initial assessment of the food systems in India and Kenya working as the method used to assess food security requirements. The GFSI is a metric that enables the ranking or benchmarking of 26 food security indicators across 113 countries. It has been developed by the Economist Intelligence Unit (EIU). It is focused on three core pillars of food security: affordability (six indicators), availability (eight indicators with five sub-indicators) and quality and safety (five indicators with nine sub-indicators); a further category is natural resources and resilience, which is an adjustment factor (seven indicators with 21 sub-indicators).[ii] The index is a quantitative and qualitative benchmarking model where the category scores are calculated from the weighted mean of underlying indicators and are scaled from 0 to 100, where 100 = most favourable. These categories are affordability, availability, quality and safety. The Centreplate Model has been developed to identify protein diversity across meal types and it is extended here to be used to guide policymakers who wish to identify improvements in protein supply (Martindale, 2017). This is achieved by association and ranking techniques that benchmark data to protein supply statistics. The Centreplate Model extends these findings to indicate NPD processes that can integrate the use of vegetables as protein sources would be favoured in enhancing protein supply. This is not straightforward because such developments would require concentrated vegetable protein as an ingredient to fortify food products. The rank analysis has been developed as the Centreplate Model for dietary policy in the UK because it reflects the importance of protein portions of meals. The model does enable the identification of pressure points and opportunities in supply chain functions with respect to specific food item categories. The importance of the protein content of foods is a major driver of dietary change and it was selected for this reason. A decreased diversity of dietary choice across national indicators highlights the requirement for innovative and flexible incentives in the food sector that establish NPD strategies that provide a wider range of products. The incentive for reducing food waste in the supply chain may well be provided by considering the limitations to NPD and diversifying the retail offer to large populations. This activity has been established for farm and product diversification in the UK and insights from these developments are of value here. Food processing and preservation have important roles to fulfil here because the development of new products (e.g. fruit juices) and stabilized ingredients (e.g. maize starches) will enable waste reduction in supply chains from producer through to consumer functions. A critical component here is to understand how to unlock data held by the suppliers and supply chain because this will stimulate changes to infrastructure and transport that are necessary for robust food security for food markets in India and Kenya. Currently, the requirement to diversify protein choice is apparent, so the ability to concentrate vegetable proteins for ingredients and stabilize vegetable products so that they can be processed into ingredients is of value. The ability to obtain more incisive supply chain data is a universal goal for improving protein supply and reducing food loss. The use of interventions for preservation techniques such as freezing and drying with storage interventions such as the use of fit-for-purpose packaging have been shown to result in less food waste in the UK (Martindale, 2014). The impact these waste-reduction models have on NPD processes has also been tested for the UK food system and the research reported here indicates that similar approaches to food and beverage NPD could be utilized in the food system for global impact.

1.7. Conclusion

The inequality of resource distribution in the global food system must be tackled if the UN SDGs are to be realized. This chapter has demonstrated how trade will change distribution of resources with respect to value and processing. This dynamic will need to continue if we are to continue to innovate, but it is the investment in innovation that decides where processing is distributed and where wealth is generated. This will change and our unequal global food system simply follows that high-income countries have more resources than low-income countries. One of the most visible demonstrators of this is shown in Fig. 1.5, which shows the number of calories wasted per person per day for different nations globally. The data are derived from an important paper written by Chen *et al.* (2020), which shows similar relationships for most nutrients; their loss is greatest in those nations that have access to nutrients and can afford to waste them. If sustainable outcomes are required, the global food system cannot seek to bring all citizens up to the level of wastage currently seen in high-income nations. There needs to be a rethinking of policy and action that brings waste to a level where there is greater equality. Fig. 1.5 demonstrates a universal principle in sustainability that is often seen where resources are distributed on unequal terms. We must understand why this happens in order to change it. The first step in that process is to map where resources are and where they are going; without this information there will be no sustainable outcome. Internet of Things applications offer a means to do just this.

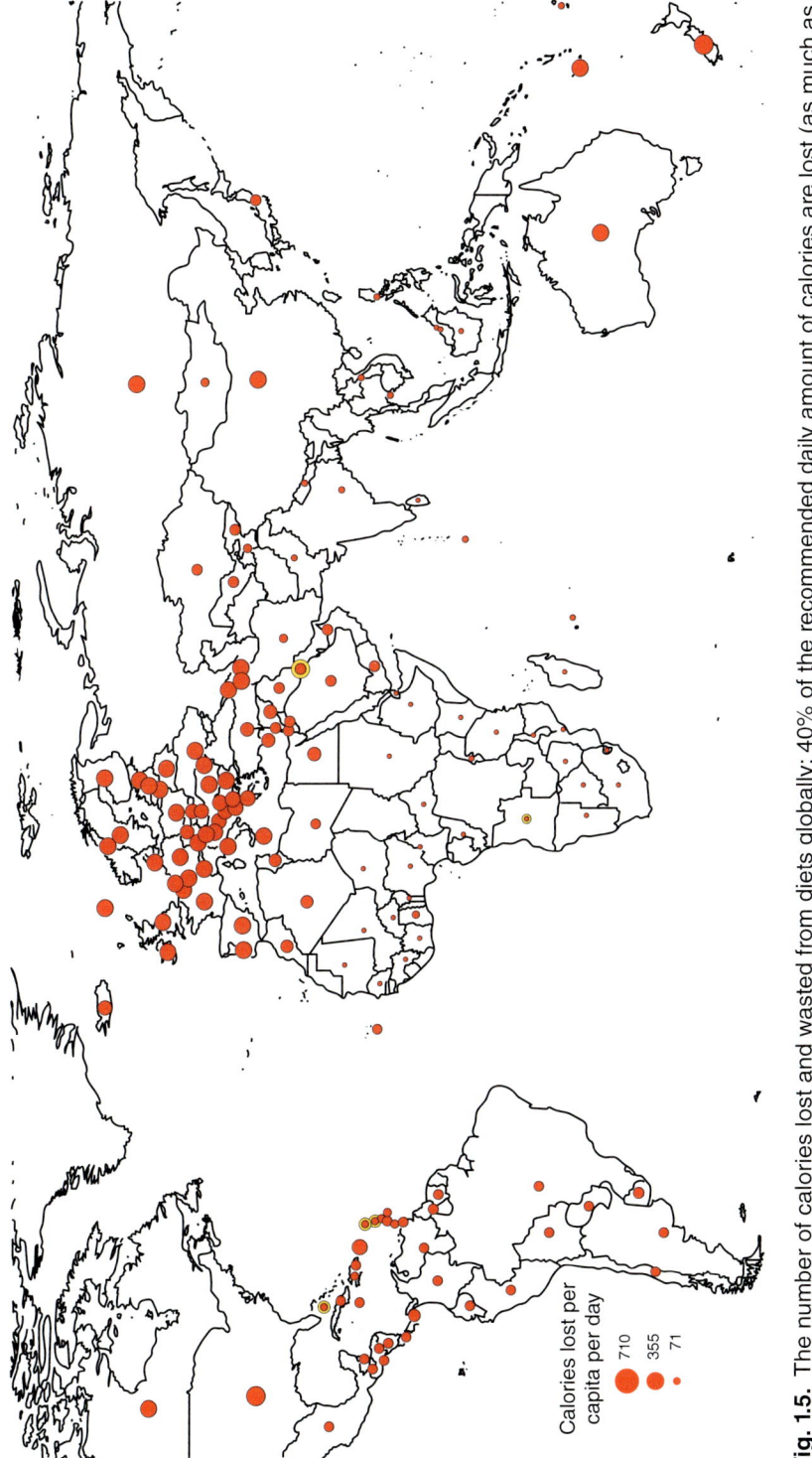

Fig. 1.5. The number of calories lost and wasted from diets globally; 40% of the recommended daily amount of calories are lost (as much as 710 calories) and the equivalent of 33 wasted diets per capita per day are lost in high-income nations compared to 4 wasted diets per capita per day in low-income nations. Wasted Daily Diets is calculated, according to Chen *et al.* (2020), as the minimum of all the Wasted Nutrient Days (24 nutrients plus energy) for a given country. Wasted Nutrient Days are calculated as the annually wasted nutrient amounts per capita divided by country average nutrient Daily Recommended Intake. (Base World Map image is the intellectual property of Esri and is used herein under licence. Copyright © 2020 Esri and its licensors. All rights reserved.)

Notes

[i]'A World Without Want', an essay written by Indira Ghandi for the *Encyclopaedia Britannica Yearbook 1974* published in 1975.
[ii]*Global Food Security Index 2018: Building resilience in the face of rising food-security risks.* © The Economist Intelligence Unit Limited 2018. See https://foodsecurityindex.eiu.com/ Resources (accessed 6 August 2019).

References

Brundtland, G.H. (1987) 'Our common future: call for action'. *Environmental Conservation* 14, 291–294.

Casini, M., Bastianoni, S., Gagliardi, F., Gigliotti, M., Riccaboni, A. and Betti, G. (2019) Sustainable development goals indicators: a methodological proposal for a multidimensional fuzzy index in the Mediterranean area. *Sustainability* 11, 1198.

Chaudhary, A., Gustafson, D. and Mathys, A. (2018) Multi-indicator sustainability assessment of global food systems. *Nature Communications* 9, 848.

Chen, P.-C., Yu, M.-M., Shih, J.-C., Chang, C.-C. and Hsu, S.-H. (2019) A reassessment of the Global Food Security Index by using a hierarchical data envelopment analysis approach. *European Journal of Operational Research* 272, 687–698.

Chen, C., Chaudhary, A. and Mathys, A. (2020) Nutritional and environmental losses embedded in global food waste. *Resources, Conservation and Recycling* 160, 104912.

Grunert, K.G., Hieke, S. and Wills, J. (2014) Sustainability labels on food products: Consumer motivation, understanding and use. *Food Policy* 44, 177–189.

Gustafson, D., Gutman, A., Leet, W., Drewnowski, A., Fanzo, J. and Ingram, J. (2016) Seven food system metrics of sustainable nutrition security. *Sustainability* 8, 196.

Haddad, L. (2018) Reward food companies for improving nutrition. *Nature* 556(7699), 19–22.

Lester, S.E., Costello, C., Rassweiler, A., Gaines, S.D. and Deacon, R. (2013) Encourage sustainability by giving credit for marine protected areas in seafood certification. *PLOS Biology* 11, e1001730.

Martindale, W. (2014) Using consumer surveys to determine food sustainability. *British Food Journal* 116, 1194–1204.

Martindale, W. (2016) The potential of food preservation to reduce food waste. *Proceedings of the Nutrition Society* 76, 28–33.

Martindale, W. (2017) Cutting through the challenge of improving the consumer experience of foods by enabling the preparation of sustainable meals and the reduction of food waste. In: Morone, P., Papendiek, F. and Tartiu, V.E. (eds) *Food Waste Reduction and Valorisation*. Springer, Cham, Switzerland, pp. 7–23.

Martindale, W., Hollands, T., Swainson, M. and Keogh, J.G. (2018a) Blockchain or bust for the food industry? *Food Science and Technology* 32, 40–45.

Martindale, W., Swainson, M., Hollands, T. and Marshall, R. (2018b) Bread winner. *Food Science and Technology* 32, 32–35.

Martindale, W., Swainson, M. and Hollands, T. (2019) New direction for NPD. *Food Science and Technology* 33, 30–33.

Martindale, W., Duong, L. and Swainson, M. (2020a) Testing the data platforms required for the 21st century food system using an industry ecosystem approach. *Science of The Total Environment* 724, 137871.

Martindale, W., Swainson, M. and Choudhary, S. (2020b) The impact of resource and nutritional resilience on the global food supply system. *Sustainability* 12, 751.

Martindale, W., Wright, I., Korir, L., Opiyo, A.M., Karanja, B., *et al.* (2020c) Framing food security and food loss statistics for incisive supply chain improvement and knowledge transfer between Kenyan, Indian and United Kingdom food manufacturers. *Emerald Open Research* 2, 12.

Noakes, M. and Clifton, P.M. (2005) *The CSIRO Total Wellbeing Diet*. Penguin, Camberwell, Victoria.

Rejeb, A., Keogh, J.G., Zailani, S., Treiblmaier, H. and Rejeb, K. (2020) Blockchain technology in the food industry: a review of potentials, challenges and future research directions. *Logistics* 4, 27.

Rueda, X., Garrett, R.D. and Lambin, E.F. (2017) Corporate investments in supply chain sustainability: selecting instruments in the agri-food industry. *Journal of Cleaner Production* 142, 2480–2492.

Sachs, J.D. (2012) From millennium development goals to sustainable development goals. *The Lancet* 379(9832), 2206–2211.

Sala, E., Mayorga, J., Bradley, D., Cabral, R.B., Atwood, T.B., *et al.* (2021) Protecting the global ocean for biodiversity, food and climate. *Nature* 592(7854), 397–402.

Wilson, M. and Jackson, P. (2016) Fairtrade bananas in the Caribbean: towards a moral economy of recognition. *Geoforum* 70, 1121.

World Commission for Environment and Development (1987) *Our Common Future*. Oxford University Press, Oxford, UK.

2 Mapping Data: New Approaches for Food System Applications

WAYNE MARTINDALE

2.1. Who is Defining the Digital Revolution for Food and Beverage Manufacturers?

Food and beverage industries can be revolutionized by the ability to increase the capacity of collecting data through the application of Internet of Things (IoT) and other digital technology frameworks. The industry itself has a legacy of being data-rich in that it is required to collect and store both process and product information in formalized auditing procedures that are focused on food safety and legal compliance. Audit procedures are used globally for food safety management systems and this chapter will consider whether it is reasonable to consider how the data for being 'audit ready' can be used to assess other aspects of business performance such as sustainability assessment. There is no doubt that audit frameworks such as BRC (British Retail Consortium) and others have made the global food system safer – they most certainly have. The question this chapter seeks to develop is as follows. These audits frameworks have been established to communicate and qualify compliance in supply chains for specific suppliers and customers. In doing so, have they become a means to qualification rather than one of continuous improvement? The question initially arose because much of the recent food safety commentary has questioned why audited business still have crises and failures of food safety procedures. Improvements in more targeted auditing approaches have highlighted this failure, for example, the use of data mining in forensic financial accounting, where irregularities are used as a potential indicator of safety crisis or criminal activity. The ability to have audit compliance and poor safety practice occurring has troubled safety practices and two areas of change management have provided solutions for this.

The first area is the issue of assessing safety or business culture within an organization – that is, how do individuals behave in response to a food safety

© CAB International 2022. *Food Industry 4.0: Unlocking Advancement Opportunities in the Food Manufacturing Sector* (W. Martindale *et al.*)
DOI: 10.1079/9781789248593.0002

management system? Many of the food crises that have emerged in a food industry that is highly regulated and audited are because of failures in the culture of organizations. Examples include sub-standard ingredients being accepted because cost challenges move individuals to procure them, temperature abuse records ignored if temperatures do not meet compliance requirements, and incorrect labelling of allergens on a product. These types of non-compliance issues are all determined by the actions of individuals who may take the wrong decision, and this is influenced by the behavioural culture in an organization. This used to be the subject of psychological research. The food and beverage industry has come to accept that behaviour of individuals in organizations has an impact of food safety and audit compliance. It is no longer a highly academic issue of organizational or behavioural culture. The culture has become a critical business process that has been shown to enhance and strengthen good practice. More importantly, the development of a strong values culture protects against crisis and provides resilience in crises associated with food safety. Recognizing behaviours and practices within an organization, with respect to values the organization holds, is as important, and it has transformed the concept of value added in manufacturing. This is because it relates to values associated with sustainability and corporate social responsibility; these have become core to business development and planning. In food and beverage manufacturing businesses it is associated with the food safety management systems of those companies, where it reinforces assurance associated with products.

The second area is driven by the new capabilities that technologies can provide, distributing trust by logging data inputs that cannot be changed once they are included into a record-keeping system. These are the distributed ledger technologies (DLTs) and blockchain platforms that have emerged from the financial technology arenas, where a system was needed to provide trust for digital currencies such as Bitcoin. These have no regulating central bank that sets value thresholds for transfers and transactions, so the need for trust is critical; the money transfer needs to be held by the vendor and it needs to be received by the customer, otherwise the value held in such a system will not work. These technologies have been identified as a means to deal with bad data if good data are placed into a supply chain in the first instance. This is because they work on the principle of mass flow in that what goes into a supply chain must come out of that supply chain. They also work on the principle of garbage in–garbage out (GIGO). If there is an irregularity, it can be traced to the source of it because blockchain systems are immutable – they cannot be changed once the data are in the chain.

Such advances in culture and technology in food and beverage businesses have begun to reform the global food system. Improvements in manufacturing over many decades still highlight gaps where crisis will occur and resilience will be compromised. The following have all improved reporting, compliance and consumer experience: (i) regulation for environmental and consumer safety; (ii) use of key performance indicators that assess environmental and sustainability; and (iii) integrating technologies that improve productivity and process efficiency. The occurrence of crisis means they need to develop further and this is recognized by regulators who are developing frameworks, such as the Food Safety Modernization Act in the USA. This seeks to integrate advances in understanding culture and technologies so that they can provide and assess the robustness of trust in supply chains. The approach is very different from what has gone before because it is now recognized many consumers will decide purchase and utilization preferences across thousands of product stock-keeping units sourced from globalized supply in a small number of shopping trips. If there were a 1% error rate in labelling across these products shown to consumers – that, is they were 99% correct – then there could easily be 100 errors that could be associated with mislabelling of allergens, for example. This highlights why the food system needs to analyse data smarter and be more incisive in response to errors. The issue of being audit-ready is in place for many manufacturers; their need to be crisis-ready is part of the modernization of the food and beverage industry. The integration of improving both culture and technology associated with assurance and safety in food companies has highlighted the use of blockchain and similar technologies in securing and transferring data.

The requirement to be audit and crisis ready is crucial in a globalized industry where there are constant changes in labour force, markets, technologies and formulations. The ability to demonstrate resilience in the event of any non-conformance or crisis is just as important now the industry has developed a platform of producing and manufacturing safe products. Resilience is an attribute that this chapter will seek to define because it is concerned with analysing and mining the large volumes of data collected by food and beverage companies in achieving audit compliance. The actual volume of data utilized by auditing procedures is thought to be low and it is called dormant or, as this chapter will prefer to call it, latent data because it is ready to be used. The actual volumes of data and audit compliance are not actually well known, even though we assume they are of large volume. A search of academic literature provides several articles for IoT frameworks, models of data structures in organizations and reviews of the semantics of the number of papers published referring to IoT.

If a food and beverage manufacturer wishes to know how much data capacity is required daily in terms of the volume of bytes collected for food and beverage manufacturing processes, it will not be easy to find. This may well surprise many practitioners as the view is the food and beverage industry is being revolutionized by IoT and digitization because there are gaps and non-compliance throughout processes and product development that need to be addressed. It is important to appreciate this before we consider how data are used in the industry because the requirement for data must be established before a consideration of what audit, what technology and what goals are required. Manufacturers are typically the source of this knowledge because they know the volume of product produced for each unit of investment in labour, materials and finance. They need to comply with safety outcomes so that products are not misreporting the nutrient declarations, not failing to manage allergens and not disregarding food safety requirements. The current IoT and digitization revolution is an attractive proposition to many manufacturers because what is very well known is the number of certification and auditing schemes that are available to them. The requirement to choose a specific audit framework or certification is typically determined by their customers, who will specify supplier requirements. How this is carried out is beyond the control of manufacturers because the ability for them to provide proof of principle is not within their expertise and auditing your own activities is not enough for supply chain compliance. A competent authority is required to provide trust for audited processes, and this is where digitization technologies are providing some promising potential, because these demonstrate how to stimulate the growth of trust between suppliers and customers.

2.2. Food Safety Management Systems, Data Integrity and the Emergent Requirement for Food Defence Against Criminality

Food safety management systems have transformed how and when data are collected in food and beverage companies beyond financial accounting. It is important to understand the similarities between reporting food safety management and sustainability data because each depends on the transfer of robust and assured information associated with processes and products. Technological advances and procedures such as blockchains can provide a method of transferring different types of information for different purposes at the same time. The issues of sustainability and safety in food supply chains are also concerned with the identification of control points, critical control points in food safety management systems. The activities at these control points are critical to any future events because if they are not controlled food safety or other outcomes such as sustainability impacts will be compromised. The events and processes embedded by companies at these control points can be bespoke and proprietary, even though they will be developed to increase food system resilience and reduce the risk of crisis. Food defence plans have become an extension to the hazard analysis critical control point (HACCP) plans because of the issue of food fraud, food contamination and wider food integrity issues. The food defence plans include threat analysis from ideologically motivated individuals or events (TACCP) and vulnerability analysis from economically motivated individuals and fraudulent (e.g. raw material fraud) events (VACCP). HACCP, TACCP and VACCP focus on different aspects of food safety and they should be kept separate so that identification of control is not lost. The same is true for sustainability assessment. All of these methods can impact on trust and brand value, where losing trust and integrity means business loss and damage. The first line of defence in any organization is cultural, involving human intervention and control. With interventions such as blockchain technology and inspections, we can deal with bad data, but these cannot readily deal with bad people. Most food safety and food defence breaches are concerned with failure of trust or culture and this can often be defined in terms of criminality (van Ruth *et al.*, 2017). The complexity for the global food system means criminality is constantly changing because of the influence of external market activities such as pressures on profit margins, sustainability awareness, application of new technologies, changing consumers, nutritional requirements, trade and origin or integrity of product. This means change and resilience in response are critical components of business success, where the following risks must be removed and, if they do occur, impacts reduced. These include chain of custody abuses, substitution of products, fraudulent harvest or slaughter, fraudulent enhancement for quality, counterfeiting, dilution or concealment, mislabelling and contamination. We have seen the importance of fraud increase and it is central to any substantiation of food integrity as well as concerns regarding bioterrorism (Masset *et al.*, 2014). The global food system faces different challenges in this century with the establishment of globalization. The food security issues faced are very different in a world population who recognizes the value of food quality alongside quantity. The influence of digital communication and social media has also changed many aspects of carrying out manufacture and business (Hsu and Lawrence, 2016).

The central messages are called the seven truths of food crime, as follows:

1. Food crime is not a food problem, it is everyone's problem.
2. Food crime is a silent threat.
3. Food does not defraud people, people defraud people.
4. Food criminals are mostly food people.
5. Food criminals are just like us.
6. Being proactive is essential.
7. There is no silver bullet.

The definition of digital revolution and the drivers for it are therefore complex. Food safety and regulatory compliance are core because the food system of the 20th century was focused on delivering safe quantities of food and beverage products to consumers. The 21st century has presented something new in that technologies that secure and analyse large volumes of supply chain data collected for safety and compliance can now mine datasets to provide quality and added value outcomes. These include reporting sustainability values and consumption impacts, where a common or universal attribute is the use of geographical information associated with a location and time when a material or product was utilized. It is the geographical information that concerns us here because it provides an important additional proof for integrity. We have already appreciated the mass flow approach to integrity in that what goes into a supply chain must come out and geographical information establishes that a material or product cannot be in two places at once. This provides an additional proof of integrity and it is one that should develop for increased assurance of consumer goods.

2.3. Demonstrating Resilience and Sustainability Using GIS and LCA Methods to Develop a Platform Digital Twin Demonstrator

This leads us to the modern view of data and resource flow in food systems where carbon footprinting and life cycle assessment (LCA) methodologies are used to determine the impact of greenhouse gas (GHG) emissions associated with crop production, food manufacture and the distribution of food and beverage products to consumers (Martindale, 2015). There is an alignment with data management practice in food and beverage manufacturing and the collection of data for food safety management systems. Both of these data flows are adding value to a business, enhancing the brand and integrity associated with practice. It is how these data streams can be utilized together that concerns us in this chapter. LCA methodologies have enabled food manufacturers to understand the impact of the amount of GHG emissions associated with the production of agricultural produce and manufactured food and beverage products. Geographical data are crucial because resources, people and products move through factories and supply chains. An important example of resource flow at a basic level is provided by water availability, where irrigation of agricultural systems is most intense where water scarcity is most acute. With many temperate zones experiencing Mediterranean and sub-tropical climate, there are emerging challenges for food manufacturers here (Gain *et al.*, 2016). Data resources are critical in determining whether information is correct or can be trusted and there are systems that can help enhance both of these, such as blockchains and DLTs, which distribute responsibility for trust (Martindale *et al.*, 2018). This is demonstrated for carbon footprint data that can provide a measure of supply chain efficiency for processing and manufacturing inputs from farm to fork or farm to taste.

The ability to bring culture and technological changes into formalized business systems has important implications for reporting information regarding the sustainability of food and beverage products. The methodology used to measure environmental impacts of processes and products is LCA, which has been demonstrated for supply chain and system-wide activities. LCA accounts for energy utilized to produce and consume products by identifying yield of a system as a functional unit, which is a specified mass of product, such as a tonne of biomass. The LCA defines and works very well within boundaries for detailed analysis. It is important to appreciate that food and beverage products move through supply chains from agricultural producers, food manufacturers to retailers and consumers; there are several operators and functions that can change in response to resource availability. This variability is what LCA methods do not deal with very well; in many respects they are not resilient, even though they deliver accurate and precise data regarding processes. This means there is a requirement to package LCA outcomes and data into a system that can map process and product resource allocation in supply chains geographically as they move through supply chains. Geographic Information Systems (GISs) do this. This chapter presents initial GIS–LCA models for food and beverage supply. The data needed to develop these GIS–LCA demonstrators have increased in terms of amounts available from government agencies, the accessibility of different types of data and the timeliness of reporting national statistical data. This has provided realistic development of Digital Twin applications that use GIS–LCA to project food system outcomes, such as identifying where the sustainable or circular economy value can be most realized.

Developing the GIS–LCA methodology means there is potential for the food and beverage system to be capable of providing instantaneous audit of resource flows and product inventory for whole supply chains. This would mean that the data regarding supply chain inventory could be aligned to the metrics described in the United Nation's Sustainable Development Goals (SDGs), because it would report the GHG emissions, food loss or resource use at any time during a production cycle to provide a transformative assessment of sustainability for fast-moving consumer goods (FMCGs). An initial example of this approach is reported in this chapter, where functions such as the amount of time a product is in a chilled or frozen environment can be recorded for its product history. This can change in response to consumer demand, where this influences time spent in refrigeration, which in turn will impact on carbon footprint. For example, when consumer demand is high, chilled foods will spend less time on refrigerated shelves in retail environments, meaning the product carbon footprint decreases. So getting the balance between production, logistics and demand just right will optimize this lowering of carbon footprint. This is dependent on as close to instantaneous data flow across supply chain operators as possible for optimal carbon footprint and waste-reduction outcomes, because operations are no longer working in a flywheel motion that cannot stop and tends to overproduce. Instead, production can respond instantaneously with demand when data flow from the demand 'downstream' or consumption functions of the supply chain can be virtually instantaneous. It has enabled the development of more robust Digital Twins that can project outcomes for system-level events across supply chains and populations, where the use of data from the national census to project dietary scenarios has already been tested.

The hybridization of LCA and GIS has been tested using logistical data with GHG emissions and social costs, using conversion factors derived from secondary data. An example of this approach is shown in Fig. 2.1, which is a demonstrator of the GIS–LCA for distributing products within 70 km of an urban centre in the UK (the Leeds area) for ten meat manufacturing companies based in the region. The spatial information for UK beef production is also shown in Fig. 2.1. It is unlikely that the relationship between farm production and food manufacturing will always be closely related because of the requirement to deliver products to urban centres. The GIS–LCA method offers the potential to define how agricultural production and food manufacturing can be planned with the delivery points. Fuel consumed has been calculated using conversion constants for product freight described in research presented by Martindale and others (Martindale *et al.*, 2008). Fuel consumed for whole freighting operations (products and vehicles) has been obtained using the UK Department of Transport Freight Best Practice KPI publications, typically 3.6 km per litre of diesel for a heavy goods vehicle (HGV) or large goods vehicle (LGV).

The conversion factors for the economic cost of food transport for each 1000 km in this study were £4.40 for GHG emissions, £31.20 for total social cost that includes accidents, £222.60 for congestion, £0.80 for transport infrastructure, £5.70 for noise and £10.10 for air quality using reported conversion factors (Smith *et al.*, 2005). The greatest costs of transporting food and beverage products in the UK are social costs that include accidents and congestion, which directs the analysis for conserving transport resources to the thorny issue of more informed planning of transport rather than a shifting of fuel sources to more carbon-neutral options for transport as a solution to a range of environmental impacts. The conversion factors are obtained in this study utilizing the costs of GHG emissions, accidents, congestion, transport infrastructure,

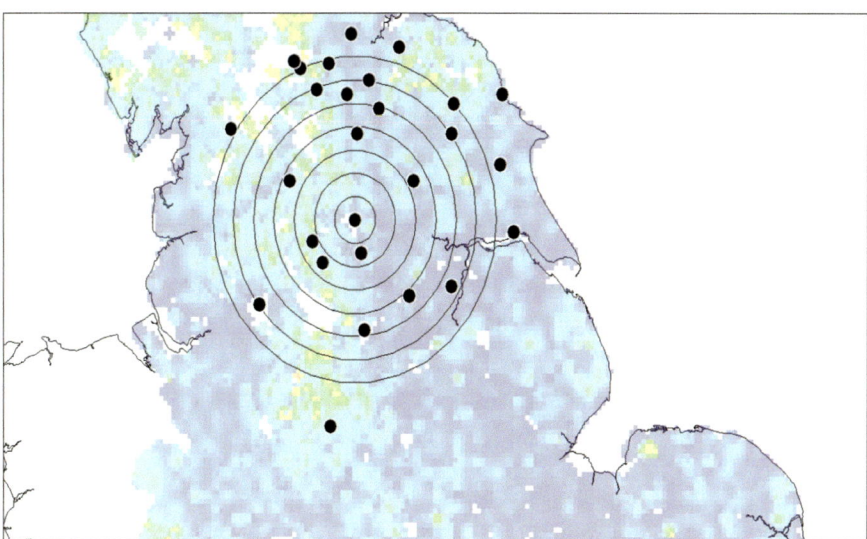

Fig. 2.1. Logistical operations and product flows for meat manufacturers. (Copyright ©
W. Martindale 2020, developed in MapInfo Pro 10.0. and the grid square agricultural census
data, as converted by Edinburgh University Data Library, are derived from data obtained
for recognized geographies from the Department of Environment, Food and Rural Affairs
(DEFRA), The Welsh Assembly Government, and The Scottish Government (formerly
SEERAD), and are covered by Crown Copyright.)
Note: This demonstrator shows the initial Digital Twins developed to assess how food
companies report carbon emissions to carbon management schemes (e.g. for carbon
credits). Retailers and customers increasingly require GHG emission information for their
products and supply chains because corporate social responsibility (CSR) outcomes are
communicated and mapping footprints can provide a method for doing this, shown here. Beef
production data have been obtained from the Agcensus database. It represents a resolution
of 2 km^2 and demonstrates the distribution destinations for ten meat manufacturers within
70 km of Leeds city centre. CSR criteria can be mapped for these points. The blue
background grid shows the intensity of beef herd production with the lighter blue through to
yellow/red grids showing higher intensity areas of beef cattle production.

Km from Leeds[a]	Km travelled	GHG emissions (kg)	Diesel consumed (l)	GHG cost (£)	Social cost (£)
10 (5)	374	526	196	1.66	102.95
20 (2)	90	393	147	0.40	24.82
30 (2)	157	125	47	0.70	43.23
40 (6)	270	231	86	1.19	74.17
50 (3)	129	107	40	0.57	35.55
60 (4)	265	251	94	1.17	72.88
70 (4)	114	271	101	0.50	31.22

[a]Number of destinations in parentheses.

noise and air quality reported by Defra, and dividing them by the reported food
miles by HGV (5.8 billion km) and LGV (4.7 billion km) to obtain the typical
cost per km for a particular impact (Smith *et al.*, 2005). The sum of accidents,

congestion, transport infrastructure, noise and air-quality cost is presented as the sum of social cost and this is sometimes called the true cost accounting method, which is relevant to resource management. This is because it provides an important insight beyond fuel costs for fiscal planning, data management and carbon footprinting that can all be integrated for meaningful applications in the food system. The limits in regional agricultural product supply have been traditionally extended by logistical infrastructure, preservation and packaging. We can account for these limits in terms of GHG emissions and social costs for sustainable business systems.

2.4. Taking Consumer Insights Further with Instantaneous Audit and Digital Twins?

The farm-to-fork view of supply chains aligns with these types of GIS–LCA methods, but a future goal will be to provide a farm-to-taste assessment that includes the functions of utilization, consumption and consumer experience (Martindale, 2017). The farm-to-taste view is system-wide and feedback on the taste, aroma, nutrition and dietary impacts – which are the consumer experience – is becoming possible using digital solutions that include systems built to efficiently collect retail sales data. The experience of foods is often left outside of the analysis boundaries of LCA, but this is some of the most important information because it determines purchase and popularity. The GHG emissions associated with them are even identified in the USA by the Environment Protection Agency as potential targets for full assessment as Scope 3 GHG emissions, which are those associated with the value chain. Until digital technologies demonstrated how DLT and blockchains could trace high-volume and high-variability financial flows in systems as immutable data, it was thought Scope 3 emissions were far too variable and complex to deal with. They are not. These types of value chain processes have a role to play in how much product is likely to be used and wasted. Their impact is significant in understanding how to project resource flows in models called Digital Twins (Martindale et al., 2020). Digitization has made the projection of these changes more incisive, where prior models could only realistically produce a projection of a likely state of change. Digital Twins are tools that utilize large datasets to project change and they are dynamic because feedback from point of sale or consumption will be continuously updated (Martindale et al., 2019). The current goal of many digital companies is to make these feedbacks instantaneous as changes in the food production or consumption actually occur and in doing so to provide a real-time inventory of resources in supply chains.

While many of the currently used blockchains have developed so that they enhance assurance and reduce the risk of safety failures, the reach of their applications goes much further in that they are secure and gate collective sources of supply chain data. This is dependent on as close to instantaneous data flow across supply chain operators as possible for optimal carbon footprint and waste-reduction outcomes, because operations are no longer working in a flywheel motion that cannot stop which tends to overproduce. Instead of this, production can respond instantaneously with demand when data flow from the demand; 'downstream' or consumption functions of the supply chain can be virtually instantaneous. It has enabled the development of more robust Digital Twins that can project outcomes for system-level events across supply chains and populations where the use of data from the national census to project dietary scenarios has already been tested. An example of these Digital Twin tools is shown in Fig. 2.2, which uses a GIS change agent-based model using nearest neighbour methods and cluster analysis to measure the connectivity of thousands of producers, processors and food businesses. The datasets that develop the Digital Twin are used to build sustainability and consumption space metrics at population or meta-levels. An important application of this type of Digital Twin system will be to define what supply and demand functions across the food system will enable carbon-zero or carbon-neutral goals to be reached and guide SDG targets.

The ability to interrogate food system data using metrics and report indices that will guide businesses and consumers is critical in light of organizations making dietary recommendations that can change the global food system. Nutritional delivery must be the ultimate boundary in the food and beverage system that should be met for every meal that is equitably delivered to every global citizen. Understanding how this can be achieved has been tested using the Delta Model developed by the Sustainable Nutrition Initiative at the Riddet Institute in New Zealand (Smith *et al.*, 2020). This is important because it provides clarity to defining whether a diet containing livestock products or only plant products is actually sustainable. Most outcomes now suggest a balanced diet is the most appropriate choice, but defining balance in diet is not straightforward so that guidance is required. This guidance will require scaling outcomes to populations and extrapolation of data, which many LCA experts get very uncomfortable with, but none the less we have tested this using benchmarking, market insights and national statistical data. Such an approach of building projections from Digital Twins at product concept stages seeks to scale the impact of new product development (NPD) at population scale and is termed meta-NPD. The impact of post-farm gate food and beverage consumption will require these meta-projections that sum the impact of all supply chain impacts at population scale because they can validate net-zero, carbon-neutral or climate-neutral GHG emission options. This Digital Twin system has shown diets with increased carbon footprint may result in less food waste compared with diets that have a lower carbon footprint because food waste outcomes are dependent on utilization of foods by consumers who tend to waste more perishable goods such as fruit and vegetables. These types of considerations are important at the level of a regional consumption space of 1.1 million consumers, where secondary data are used to calculate GHG emissions of food categories (Wallén *et al.*, 2004). The use of secondary data means when food categories did not align a closest match was made, and the food categories were supplied fresh in the first tests shown discussed here. It has omitted the value of the preservation, packaging and utilization of foods and beverages, which is important because a consideration of the frozen food supply chain

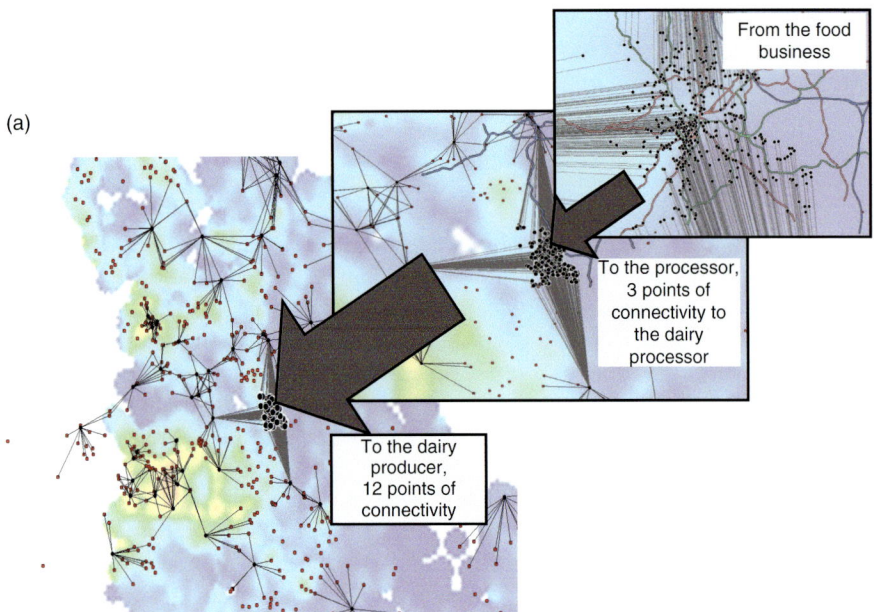

Fig. 2.2. (a) A Digital Twin demonstrator for the dairy supply chain and (b) orchard (stone) fruit in England (principally apples for sale and cider production). (a) The three node connections of 823 food and beverage retailers to 166 dairies and 202 dairy producers. The top-level data grids show the connectivity of food business reported as food business operators (FBOs) by the Food Standards Agency for England and Wales to processors; it is the three closest processors to 830 retail FBOs. The lowest-level data grid shows the intensity of dairy production derived from number of cows, where the red dots show major dairy producers and the black dots are processors. These data are from Companies House and the Agricultural and Horticultural Survey. (b) The same scenario for food businesses and customers (n = 1636) to pack house (n = 71) to processor (n = 245) and producers for orchard fruit (215). The Digital Twin supports and enables traceability back to most probable producer and processor. (Copyright © W. Martindale 2020, developed in MapInfo Pro 10.0. and the grid square agricultural census data, as converted by Edinburgh University Data Library, are derived from data obtained for recognized geographies from the Department of Environment, Food and Rural Affairs (DEFRA), The Welsh Assembly Government, and The Scottish Government (formerly SEERAD), and are covered by Crown Copyright.)

demonstrates the principle that extending shelf-life results in decreased food waste and improved sustainability outcomes (Martindale and Schiebel, 2017). The cold food supply chain reduces the risk and variability in production (e.g. seasonality) and manufacturing outputs (e.g. localizing or shortening supply chains) so that post-harvest loss and food waste due to variability in price or demand are minimized. Understanding preference and why other food groups, including bread and prepared foods, are 'thrown and not frozen' is a challenge for the food system, where shelf-life can determine the probability that foods are wasted. How preservation and packaging choice relate to the utilization of foods and cooking them is an important aspect of how sustainable they will be, which is recognized in innovative approaches to NPD.

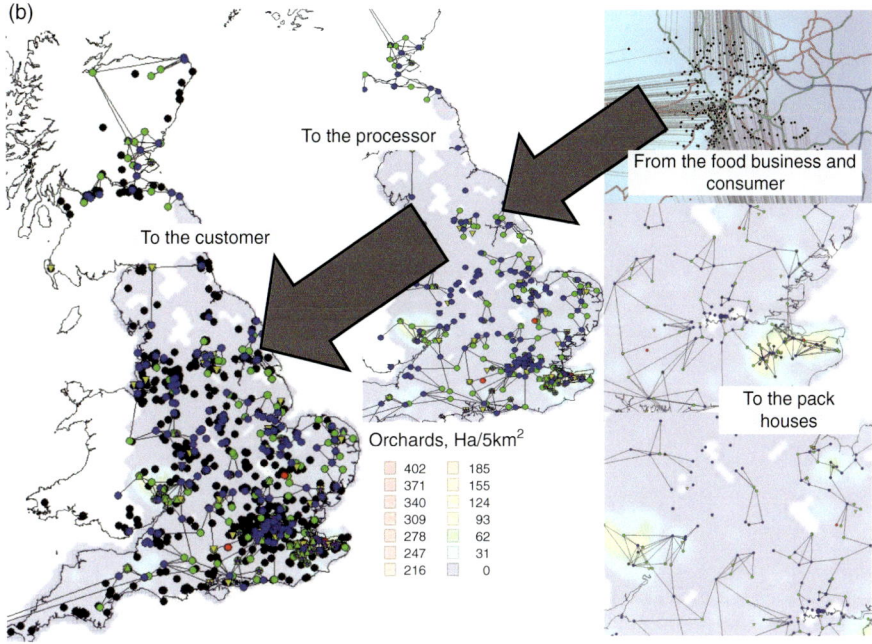

(b)

To the processor

From the food business and consumer

To the customer

To the pack houses

Orchards, Ha/5km²

402	185
371	155
340	124
309	93
278	62
247	31
216	0

Fig. 2.2. Continued.

The consumption of a more sustainable diet containing 5% less meat and 10% more fresh fruit and vegetables was assessed, as demonstrated by research from the World Wildlife Fund and the UK's Rowett Institute (Macdiarmid, 2013; Macdiarmid *et al.*, 2012). This diet was benchmarked against the National Dietary and Nutrition Survey diet for the UK, where food choices were extrapolated from the Livewell Diet and the consumption of both diets for seven days was mapped across the selected population of the Sheffield City region in the UK. The GWP (Global Warming Potential, or the GHG 'footprint') data used for food are now well established and the GIS–LCA hybrid approach in the Digital Twin allows the mapping of established LCA data with demographic data trends (Wallén *et al.*, 2004). Figure 2.3 shows the demonstrator and how national census datasets and National Dietary and Nutrition Survey (NDNS) in the UK have been used to do this for annual consumption of foods. It is important to note that beverages are not included in these diet projections. The NDNS dietary scheme is projected for a meal plan that aligns to the NDNS data, Eatwell Plate and follows the Livewell Diet, and seeks a reduced GHG emission outcome, increasing or decreasing specific food groups where necessary so the food consumed aligns with the Livewell Diet. The consumption footprint for the population investigated results in a GHG emission of 1.1 million tonnes from a population of 1.3 million people. The Digital Twin for Consumption (DTC) shows 1.1 million tonnes of GHG produced each year for consuming the NDNS diet for 1.3 million people in the region analysed, which is reduced by 5–10% if the Livewell Diet is consumed. If this level of consumption is scaled to 67 million UK consumers, the NDNS diet (the typical UK diet)

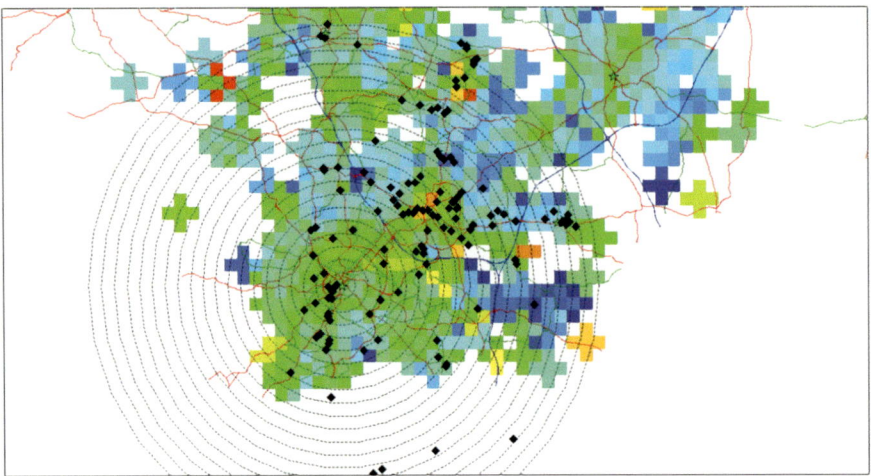

Fig. 2.3. Mapping consumption using GIS allows us to begin to present the data derived from 100 to 1000s of supply chains and build a Digital Twin that can project sustainable production and consumption outcomes, including carbon-neutral deliverables. The map generated here by a GIS shows the carbon and food waste footprint for the Sheffield City region, UK, with the population analysed at lower layer Super Output Area population for the national census, where each area has approximately 1500 citizens. The Livewell diet and typical UK citizen diet as reported by the National Dietary and Nutrition Survey were constructed for one week of consumption and projected here. The consumption spaces are 0.5 km^2. Red grids account for 30 t GHG per 0.5 km^2 and blue grids 20 t GHG per 0.5 km^2. The demonstrator shown here was benchmarked against national inventories of GHG emissions and retail expenditure so that the calibration of the Digital Twin was tested. This test demonstrated a robust projection of consumption is delivered using these methods. (Copyright © W. Martindale 2020, developed in MapInfo Pro 10.0. Contains National Statistics data © Crown copyright and database right 2012. Contains Ordnance Survey data © Crown copyright and database right 2012.)

has a carbon footprint of 54.85 million tonnes of GHG emissions, which aligns with the UK national GHG inventory for the whole agri-food supply chain, which is 55 million tonnes of GHG. Initial tests of the DTC show that diets resulting in lower GHG emissions will result in up to 25% more domestic food waste because they include greater amounts of fruit and vegetables, which are at more risk of being wasted in the domestic household. These scenarios need further testing, but the trade-offs between GHG emissions and waste production must be defined for food and beverage systems.

The initial DTC shown provides a robust test for our GIS–LCA scenario-building model using the UK national GHG inventory of 550 million tonnes of GHG. This Digital Twin framework has been used to map and calculate expenditure on purchased food from retailing. It has been shown to be robust when scaled to the national population of 67 million UK citizens, when it projects an expenditure of £144 billion each year; the total expenditure and revenues from retail, wholesale and service is £163 billion annually. While purchasing revenues are variable, the Digital Twin developed would indicate the method is robust and can be developed further. An important goal of determining the consumption footprint of populations is to guide a net-zero GHG emission consumption space by demonstrating how it will be most efficiently achieved. What is apparent from the established practice of LCA in the food and beverage sector is GHG emissions are more incisively managed in the processing, manufacturing, distribution and retailing functions of supply chains. This is because resources are metered and controlled across these functions, whereas the agri-production and consumption functions are not as finely controlled or metered. In the case of agriculture, the weather confounds attempts to measure and quantify GHG emissions, and consumption operates in arenas of consumer choice that introduce increased variability. The targets for reaching carbon-zero without offsetting or carbon crediting will require incisive methods of measuring the processes that result in GHG emissions across the food supply chain and this will require an index that can be used across food and beverage businesses.

Figure 2.2 has developed the Digital Twin to include measures of connectivity between operators in food supply chains. These are currently being tested and demonstrated for assessing food safety risks and food defence policies. The demonstrators shown are for the dairy and fruit sectors. Their application offers much promise in identifying potential risk, resource flow dynamics and most probable sources of resources in supply chains (Martindale *et al.*, 2020). Connectivity has been assessed using geographic information and regional scales for policy development and resource planning (Fernandez-Mena *et al.*, 2020; Lin *et al.*, 2019). Without engaging all food businesses, a sustainable food system will not be possible, so measurement of sustainability needs to be embedded in the culture of an industry. A sustainability index scorecard approach can help to achieve this. Testing different index approaches that use environmental impact and dietary value has identified data sources that can be collected by food and beverage manufacturers. The universal use of such measurements would be important because the use of consumer goods is interwoven into all of the 17 SDGs. While consumer goods improve the lives of billions of people, an inescapable outcome of their manufacture is increased consumption and utilization of natural resources (Martindale, 2017). As an example, in the UK there are over 350,000 food businesses. The goal of the Sustainability Index is to be applied to improve security and sustainability in each of these food businesses when products are at concept, development or manufacturing stages. This approach was developed as a model that used six attributes or functions that defined protein content of foods, distribution of products, energy embodied in their production and processing, waste associated with their use and GHG emissions (Martindale, 2017). What was specifically important in these models was the use of meal concepts developed by chefs. The method of categorization of meal types was derived from chefs and cooking books (one in particular by Jamie Oliver, *Jamie's Ministry of Food* (2008), which arranges meal types in a way that is robust and accessible).

2.5. Conclusion: Balancing a Global Diet that Connects 9 Billion Consumers

Carbon-zero thinking has been transformative in breaking the deadlock between carbon footprint and nutrition. The launch of branded zero-carbon livestock products such as whole milk, beef and lamb has shown that food producers and manufacturers are confident in making this claim.[i] The subsequent rethinking of carbon footprinting is enlightening because it can be related to achievable and nutritious diets and lifestyles so that responsible consumption is possible. It is important that improvements do not get lost in purely carbon-footprinting diets. We are developing models for the UK that identify where critical points and connectivity in the food system control resource flows (Martindale et al., 2020). NPD is the operational activity we are focusing on because if product developers and technologists build-in sustainability at the concept stages, there is an increased possibility that the final product will deliver it (Jagtap and Duong, 2019). We now have a greater understanding of how genes and metabolism interact with what we choose to eat. It is essential to keep the food system lens, and this is what the Sustainable Nutrition Initiative's Delta model does (Smith et al., 2020). Connecting datasets and making sure we speak to each other is becoming increasingly important. This is otherwise known as interoperability. In the digital arenas, we have the capability to deliver a net-zero sustainable food system, but without interoperability it will not happen. Our food future depends on all partners in the global system connecting methods and data that will guide sustainable dietary choices. The use of geographic information looks to transform the auditing arena in food and beverage because the data environment can be secured by blockchain approaches where inputs are immutable. The development of analytical and regulatory advances will lend robustness to this approach and in doing so open further applications beyond those considered here. One such area is the use of mapping applications to assess how resources, utensils and equipment move around production spaces and factories. The use of this approach in assessing compliance with hygiene regimes is already being investigated. In light of the COVID-19 crisis, defining effective hygiene control and maintenance has become crucial. Automating these processes releases financial pressures on the auditing frameworks and this means that compliance is at less risk of being compromised: most non-compliance is driven through ignorance or financial targets. Digitization enables operations to reduce the risk in this space and it is yet to be seen what impact such digital applications will have. The potential is demonstrated. Perfect products and perfect meals mean sustainable business outcomes are possible if we build them into development and cultural processes in the food system.

Note

'CN30 overview, see: https://www.mla.com.au/research-and-development/Environment-sustainability/carbon-neutral-2030-rd/cn30/ (accessed 18 December 2020).

References

Fernandez-Mena, H., MacDonald, G.K., Pellerin, S. and Nesme, T. (2020) Co-benefits and trade-offs from agro-food system redesign for circularity: a case study with the FAN agent-based model. *Frontiers in Sustainable Food Systems* 4, 41.

Gain, A.K., Giupponi, C. and Wada, Y. (2016) Measuring global water security towards sustainable development goals. *Environmental Research Letters* 11, 124015.

Hsu, L. and Lawrence, B. (2016) The role of social media and brand equity during a product recall crisis: a shareholder value perspective. *International Journal of Research in Marketing* 33, 59–77.

Jagtap, S. and Duong, L.N.K. (2019) Improving the new product development using big data: a case study of a food company. *British Food Journal* 121, 2835–2848.

Lin, X., Ruess, P.J., Marston, L. and Konar, M. (2019) Food flows between counties in the United States. *Environmental Research Letters* 14, 84011.

Macdiarmid, J.I. (2013) Is a healthy diet an environmentally sustainable diet? *Proceedings of the Nutrition Society* 72, 13–20.

Macdiarmid, J.I., Kyle, J., Horgan, G.W., Loe, J., Fyfe, C., Johnstone, A. and McNeill, G. (2012) Sustainable diets for the future: can we contribute to reducing greenhouse gas emissions by eating a healthy diet? *The American Journal of Clinical Nutrition* 96, 632–639.

Martindale, W. (2015) *Global Food Security and Supply*. John Wiley & Sons, Chichester, UK.

Martindale, W. (2017) Cutting through the challenge of improving the consumer experience of foods by enabling the preparation of sustainable meals and the reduction of food waste. In: Morone, P., Papendiek, F. and Tartiu, V.E. (eds) *Food Waste Reduction and Valorisation*. Springer, Cham, Switzerland, pp. 7–23.

Martindale, W. and Schiebel, W. (2017) The impact of food preservation on food waste. *British Food Journal* 119, 2510–2518.

Martindale, W., McGloin, R., Jones, M. and Barlow, P. (2008) The carbon dioxide emission footprint of food products and their application in the food system. *Aspects of Applied Biology* 86, 55–60.

Martindale, W., Hollands, T., Swainson, M. and Keogh, J.G. (2018) Blockchain or bust for the food industry? *Food Science and Technology* 32, 40–45.

Martindale, W., Swainson, M. and Hollands, T. (2019) New direction for NPD. *Food Science and Technology* 33, 30–33.

Martindale, W., Duong, L. and Swainson, M. (2020) Testing the data platforms required for the 21st century food system using an industry ecosystem approach. *Science of The Total Environment* 724, 137871.

Masset, G., Soler, L.G., Vieux, F. and Darmon, N. (2014) Identifying sustainable foods: the relationship between environmental impact, nutritional quality, and prices of foods representative of the French diet. *Journal of the Academy of Nutrition and Dietetics* 114, 862–869.

Oliver, J. (2008) *Jamie's Ministry of Food*. Michael Joseph, London.

Rejeb, A., Keogh, J.G., Zailani, S., Treiblmaier, H. and Rejeb, K. (2020) Blockchain technology in the food industry: a review of potentials, challenges and future research directions. *Logistics* 4, 27.

Smith, A., Watkiss, P., Tweddle, G., McKinnon, A., Browne, M. *et al.* (2005) *The Validity of Food Miles as an Indicator of Sustainable Development: Final Report.* Report ED50254, Issue 7. AEA Technology Environment, Didcot, UK.

Smith, N., Fletcher, A., Finer, O. and McNabb, W. (2020) Sustainable nutrition initiative: feed our future. *Food New Zealand* 20, 36.

van Ruth, S.M., Huisman, W. and Luning, P.A. (2017) Food fraud vulnerability and its key factors. *Trends in Food Science & Technology* 67, 70–75.

Wallén, A., Brandt, N. and Wennersten, R. (2004) Does the Swedish consumer's choice of food influence greenhouse gas emissions? *Environmental Science & Policy* 7, 525–535.

3 The Perfect Meal

WAYNE MARTINDALE

The perfect meal is the goal of providing the appearance, aromas and taste consumers expect. Everyone's view of this is different, so it is not an easily solved task, but a scientific understanding of oven cooking of meat, bread and cakes can now provide a set of perfect parameters for that perfect bake or roast. It could change the approach to new product development and responsible sourcing of food and beverage ingredients.

3.1. From Farm to Taste: A Step Further for Farm-to-fork Frameworks

There are many examples of nutrients that impact on the quality of ingredients from agriculture. Their influence in nutrition and taste has been characterized for many crops and livestock products, in particular micronutrients such as zinc, selenium, vitamin A and unsaturated oils (Bouis et al., 2019). The essential plant nutrients also do not include elements that provide functional integrity. For example, silicon is critical in providing structural support in plants and this is of importance when considering the harvesting of plants in protected or hydroponic production, where silicon may be limited (Song et al., 2016).

This is demonstrated for carbon footprint data that can provide a measure of supply chain efficiency for processing and manufacturing inputs from farm to fork or farm to taste. The carbon footprint is an appropriate means to report energy use and resource flows for many food and beverage products and it can be used to communicate impact across supply chain functions including

Corresponding author: wmartindale@lincoln.ac.uk

© CAB International 2022. *Food Industry 4.0: Unlocking Advancement Opportunities in the Food Manufacturing Sector* (W. Martindale *et al.*)
DOI: 10.1079/9781789248593.0003

consumption. This is because there are specific data inputs for each activity associated with a footprint. As an example, a typical 200 g mixed livestock and plant ingredient sandwich will have 220–290 g of GHG emissions associated with growing and processing its ingredients. Transport and packaging will contribute 20–50 g GHG emissions. GHG emissions such as methane (from livestock production) and nitrous oxide (from organic and mineral nitrogenous fertilizer use) can significantly increase these emissions. Furthermore, they can be reduced by fit-for-purpose agronomic management and efficient supply chain planning, which will be dependent on efficient data management (Martindale *et al.*, 2018).

The farm-to-fork view of supply chains aligns with these types of GIS–LCA methods, but a future goal will be to provide a farm-to-taste assessment that includes the functions of utilization, consumption and consumer experience (Martindale, 2017a). The farm-to-taste view is system-wide and feedback regarding the taste, aroma, nutrition and dietary impacts – which are the consumer experience – is becoming possible using digital solutions that include systems built to efficiently collect retail sales data. The experience of foods is often left outside of the analysis boundaries of LCA but they are some of the most important because they determine purchase and popularity. The GHG emissions associated with them are even identified in the USA by the Environment Protection Agency as potential targets for full assessment as Scope 3 GHG emissions, which are those associated with the value chain. Until digital technologies demonstrated how distributed ledger technology (DLT) and blockchains could trace high-volume and high-variability financial flows in systems as immutable data, it was thought that Scope 3 emissions were far too variable and complex to deal with. They are not, and these types of value chain processes have a role to play in how much product is likely to be used and wasted, so their impact is significant in understanding how to project resource flows in models called Digital Twins (Martindale *et al.*, 2020a).

What is most important to observe here is that the inventory of value chain impacts is well understood by successful retail businesses in the food and beverage industry because the value chain can determine product choice, preference and sales. An indication of the importance of taste and consumer experience is provided by the global production of herbs and spices reported by FAOStat, which are selected because they are used to formulate taste. Garlic production has reached 28.5 million tonnes, increasing 16%, year-on-year, since 1990; cinnamon production reached 0.22 million tonnes, increasing 12%, year-on-year, since 1990; and vanilla, which is more complex, reached 0.08 million tonnes produced globally, increasing 8%, year-on-year, over this period, which saw the emergence and growth of globalization that changed consumption dramatically. Cereal and sugar crops such as wheat and sugar cane generally show a less than 5% year-on-year increase over this period and tend to have constant rates of production in developed and regulated markets that import agri-products such as spices for taste. Removing farm-to-taste approaches from any system model will not provide a full account of what is happening with resources in the food system (Martindale, 2017b). Digitalization has made the projection of these changes more incisive where prior models could only produce a projection of a likely state of change. Digital Twins are tools that utilize large datasets to project change and they are dynamic because feedback from point of sale or consumption will be continuously updated (Martindale *et al.*, 2019). The current goal of many digital companies is to make these feedbacks instantaneous as changes in food production or consumption actually occur, and in doing so provide a real-time inventory of resources in supply chains.

3.2. Why Should the Cooking of Products Concern Us for Industry 4.0?

How fried, boiled, baked or roasted foods appear and smell when they are placed on the plate is important because this can help to define how fulfilling a meal will be. The caramelized surface of the roasted meat and vegetables, for example, the very golden colours of bakes, and home-style aromas are often foundation marketing techniques. The aroma of baking in stores, restaurants and houses can stimulate the purchase of everything from bread to real estate! This is, of course, an idealized view – that we follow the aroma of the bakery or rotisserie hypnotically in search of that perfect food that is producing them. While part myth, a consumer view of home-style cooking is important because even a small group of people will show a wide range of preferences. When a perfect cooking technique delivers the expected result for a wide range of preference, fulfilment is delivered for marketplaces and populations. This approach is extending our work in NPD kitchens and laboratories to meta-NPD outcomes, and it has a positive impact on dietary and sustainability outcomes. The starting point is to develop the cooking concept for the perfect meal and to build in consumer fulfilment to initial NPD as a concept-to-consumer process. While this all seems obvious, it is overlooked in terms of the perfect meal outcome because the service sector in the UK prepares 50% of the food purchased. The perfect meal is a must, and connecting cooking research to consumption impact can tackle some of the most important issues facing the food industry, such as reducing meat consumption, improving food safety and reducing food waste. If meals are not cooked well enough, these things will be compromised, but research continues to highlight these problems rather than recognizing the solutions that good cooking provides. Cooking processes in development and domestic kitchens are where data can be effectively collected, and it is therefore a focus for IoT applications. Most recipes are not that complex, with eight to ten ingredients from which are derived thousands of different flavour and fragrance combinations. The use of data-filtering and -association techniques for recipes is well developed so that the database structures for product development are defined. The association and filtering of these recipe components is another database issue. This is where sensory responses are collected and there is a multitude of possible IoT applications. When processes are scaled from kitchen to factory or pilot plant, there is yet another layer of data collection that can be made far more efficient with the use of IoT applications.

Research that defines the outcomes of cooking foods perfectly is rarely considered in sustainability research even though it is an essential part of the food system – typically, all food needs to be processed and cooked in some form. The reasons for this are most obviously based on food safety, food stabilization and consumer fulfilment of food and beverage products. The food and beverage industry is familiar with flavour formulation and sensory testing using tasting and consumer panels but these methods rarely extend into the sustainability arena. The development of meals is yet another aspect of cooking foods perfectly that is not fully considered here. Even though products are individually tested, they are not integrated into the utilization of them as meals or diets. These gaps in the integration of disciplines with sustainability are further overlooked by the nutritional community who will often pay scant attention to how meals are developed and consumed, instead keeping a focus on specific nutrients or ingredients of concern such as salt or sugar. There are important reasons for this lack of integration, not least the fact that food utilization by consumers is complex, with millions of choices and combinations. This results in a separation of consumer fulfilment of food and food system impacts, and this the major focus of food insights research, which seeks to bring both capabilities of branding and consumer insights firmly into the sustainability arena. There are significant frameworks for starting this process of defining sustainability through branding and consumer insights, with the impact of cooking (Garnett, 2016) and even the impact of meals, including those that are 'ready to prepare' (Rivera and Azapagic, 2019) or 'ready to eat' (Espinoza-Orias and Azapagic, 2017). This should not surprise us because chefs and culinary professionals categorize meals and processes, so there is an opportunity to integrate this knowledge with existing LCA and sustainability methods that measure the impact of producing and consuming meals.

As highlighted in Chapter 1, indices of food security and sustainability identify high-level policy risks for the production of agricultural commodities and do not really provide detailed insight for manufacturing, distribution and retailing functions of supply chains or consumption of meals. This is an oversight with respect to food insights because many of the issues associated with poor nutrition or sustainability are focused on the formulation of meals. Branding of food products will regularly make the connection between products and meals, highlighting fulfilment and convenience, but sustainability outcomes such as reduced waste and efficient cooking are not realized in these relationships. It is here where research has been emergent and has identified how product development and knowledge of food utilization (i.e. making meals) can reduce food waste. It identifies links to sustainability outcomes (Martindale, 2016) and considers how these change with different meal categories (Martindale, 2017a). The culinary development of recipes has impacted sustainability because the use of specific ingredients has been researched with respect to the most utilized combinations of ingredients. The research has utilized recipe databases that span several hundred years of recipes in cookbooks and applies evolutionary fitness modelling to identify successful recipes (Kinouchi *et al.*, 2008). This approach has seen a revival, with the use of ingredient data in voice-operated devices used in kitchens to guide domestic cooking and ingredient use across world cuisines, and is highly conserved in recipes that survive for a longer time (Tuwani *et al.*, 2019). Recipes, typically, have between seven and ten distinct ingredients. Complex recipes do not survive for long, whereas convenient recipes with fewer ingredients meet the fulfilment criteria of consumers.

3.3. From Development Kitchens to the Consumer: How do We Get to the Perfect Meal?

We now have a defined route to perfect cooking and we know that the type of oven used for baking and roasting has an important impact on the food being cooked. While this has been known intuitively by chefs and cooks since we started cooking in ovens, we have not really questioned why the perfect result happens; it just happens like that. Of course, chefs will say it is because of a mystical skill, and in some ways it is, only the skill is more important than the mystical. We now have a defined focus for the processes that make the perfect cook; it is the Maillard reactions and the heat transfer. For the perfect result every time, know the oven because we can always compensate to get the perfect mix of Maillard compounds and cooking through. However, meat cuts are the tricky part of kitchen culinology because the cooking through can be, in effect, arrested by the higher oxygen environment of convection ovens. For the most robust and assured way of getting large cuts of high protein or starchy food cooked through, a radiant oven will help. The outcome is perfect cooking, less waste and improved fulfilment, all of which have very clear sustainability outcomes.

Of course, the utilization of food and beverage reflects cultural changes, adding to the complexity of ingredient choice and meal development. Fig. 3.1 demonstrates this complexity of food supply with the three main global cereal crops, two of which supply the primary protein source for most nations (wheat and rice). It uses a synteny approach to connect production to outcomes, export, food and feed utilization in this case. It emphasizes that we are not only considering foods in the global system but also feeds to produce livestock products (maize as a principal example in Fig. 3.1). Feeds are also a critical part of the food system, producing livestock products, and maize is the principal feed crop for many nations. The footprint of feed in our perfect meal is another aspect of sustainability impact that is overlooked and these layers of complexity create large volumes of data because of the billions of combinations and choices that are made possible in developing diets. This is where an appreciation for the role of digitalization in stimulating Industry 4.0 and the use of integrated digital technologies (IDTs) that capture and assess data are both now accessible and affordable for companies and consumers.

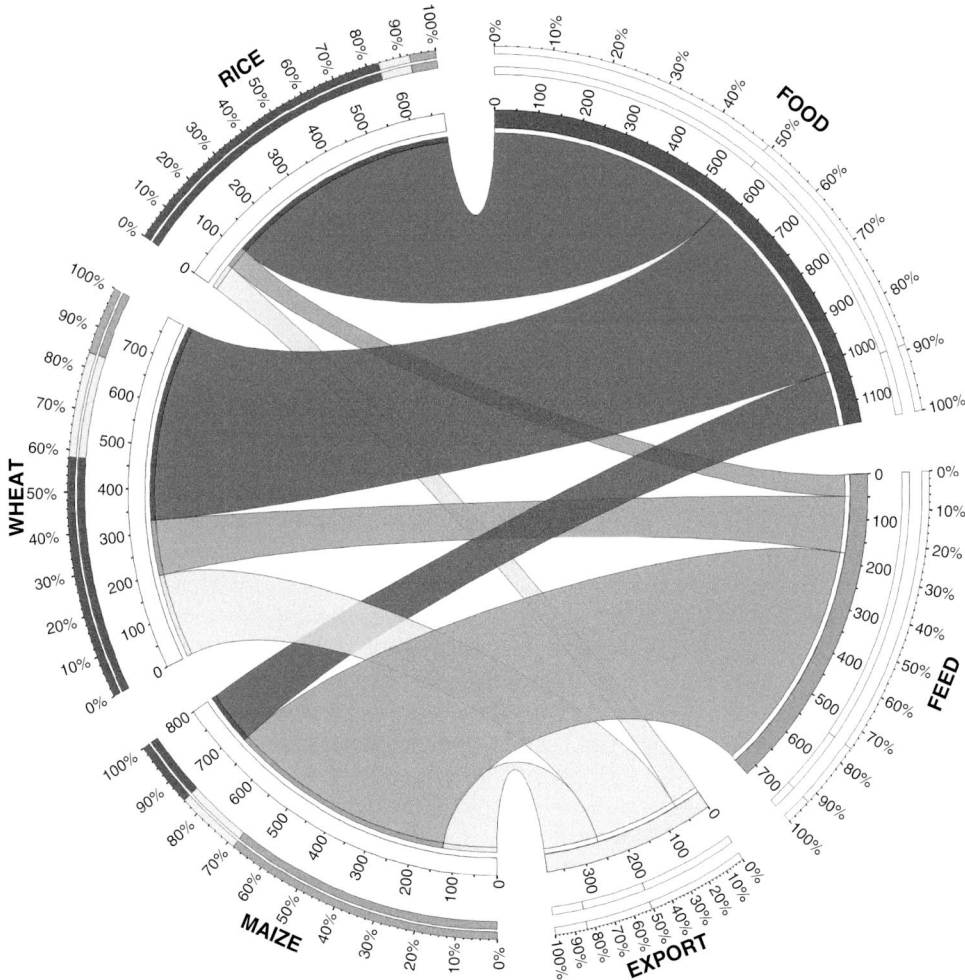

Fig. 3.1. A synteny assessment for rice, wheat and maize with food, feed and export mass (10^6 tonnes and ratio as a percentage of the total) using data from FAOStat 2013. The inner axis shows the amount of crop in million tonnes.yr^{-1} and the outer axis shows the proportion as a percentage. (Plots developed using Circos Table Viewer v0.63-9 © 2008–2019 (Krzywinski *et al*., 2009); figure derived from Martindale *et al*., 2020b)

The use of IDTs as sensors or mobile reporting devices such as mobile phones has given the requirement to consider the development of Industry 5.0 where semantic web technologies will enable consumers to interrogate and reason with the large volumes of collected data.

The development of trust is identified as a crucial element in establishing the use of semantic web technologies where ethical communication with customers in business-to-business transactions and business-to-consumer interactions is increasingly required. Methods of distributed or shared leadership and trust had been largely theoretical before the development of digitalization of data flows and the use of semantic web technologies (Alghamdi *et al.*, 2017). Food system trust will require measurement that considers both quantitative and qualitative outcomes and such indices will be transformative and have potential for scaling up to populations because of the use of digital technologies. The synteny analysis approach in Fig. 3.1 is developed for the major crop commodities that are processed globally (Fig. 3.2) for sugar crops, oil palm and soybean. Fig. 3.2 demonstrates the importance of considering how primary processing steps impact on our food system and dietary choices. Soybean is often considered a feed crop but the primary value of soybean is as an oil crop,

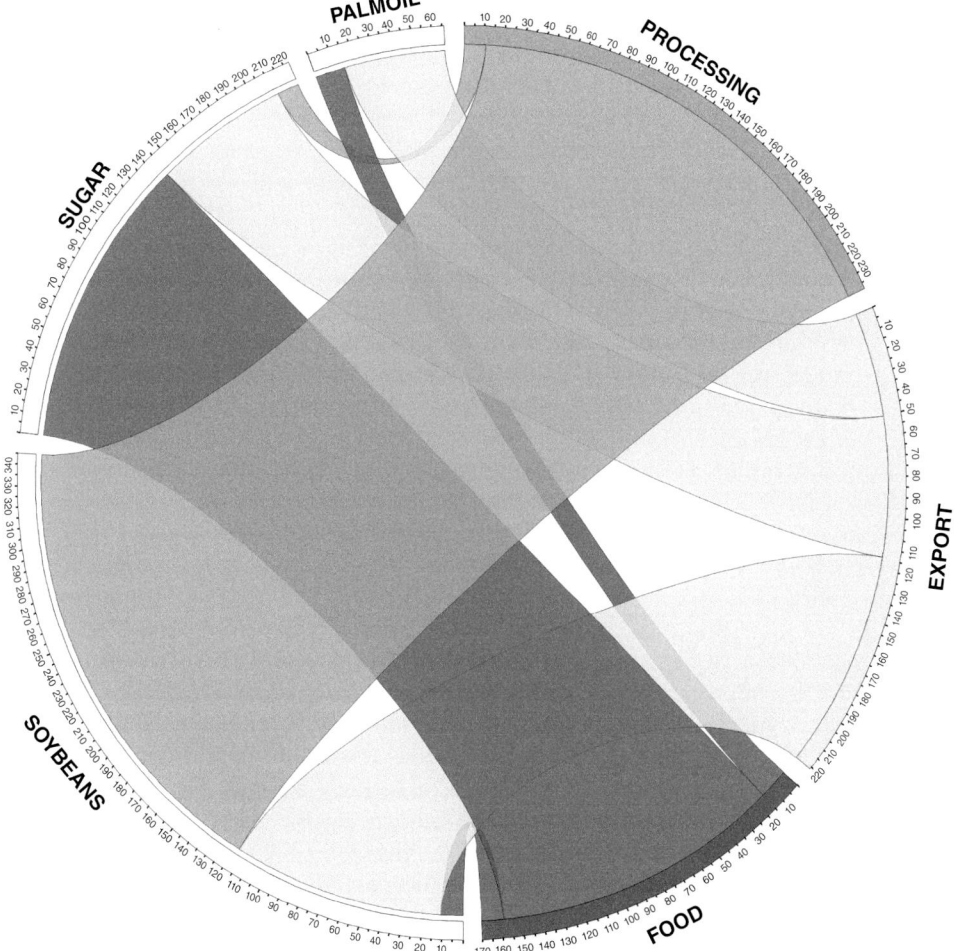

Fig. 3.2. A synteny assessment for palm oil, sugar and soybean with food, processing and export mass (10⁶ tonnes and ratio as a percentage of the total) using data from FAOStat 2013. The inner axis shows the amount of crop in million tonnes.yr⁻¹ and the outer axis shows the proportion as a percentage. (Plots developed using Circos Table Viewer v0.63-9 © 2008–2019 (Krzywinski *et al.*, 2009); figure derived from Martindale *et al.*, 2020b)

which is processed with the soy meal derived from pressing and crushing for animal feeds and high-value products. Both synteny diagrams infer the complexity in our food system with respect to what consumers utilize as food for meals and diets. It is not a straightforward process of associated oil palm with foods or soybeans with feed, for example, and requires a structured database that can provide specific descriptions of food and beverages. The food system has this framework with the UN Codex Alimentarius (Price and WHO, 2020) and the EFSA Food Composition Database (Finglas *et al.*, 2014). IoT and Industry 4.0 applications are driving the use of composition databases into a more structured format that requires the development of ontologies, so there is universal understanding of what foods are and how they are used (Rejeb *et al.*, 2022). The establishment of ontologies for food will begin the process of developing semantic web technologies that enable users to question and reason data structures. But before all this can happen, robust ontologies such as the FoodOn ontology are being developed (Dooley *et al.*, 2018). These are exciting step-changers because they link agricultural production, traded commodities and manufactured foods with the ingredients and foods consumers utilize to produce meals and diets from farm to fork (Griffiths *et al.*, 2016).

The synteny analysis presented in Figs 3.1 and 3.2 are crucial because they influence what materials consumers will utilize to create meals and diets and this means that methods of analysis are required to identify how these materials are supplied. This has been done to effectively benchmark the performance of different food categories using a simple rank-and-score analysis. The methodology has been tested and developed as the Centreplate Model which operates at the scale of diets rather than national outputs and commodity trade so that it considers the realistic protein diversification of diets (see Table 3.1). The Centreplate Model is far from the complete analysis of the perfect meal but it does start the analysis using nationally reported statistics from the FAOStat databases. It can be used to identify protein diversity opportunities across meal types and this can guide policymakers who wish to improve protein supply (Martindale, 2017a). The Centreplate Model uses association and ranking techniques that benchmark data to protein supply statistics of the food categories used by the FAOStat database. The approach has taken account of what fast-moving consumer goods companies do so well; they fully understand what drives consumers to use their products. These companies build affordability, assurance and availability into these systems and it is time to put sustainability and responsibility into them. The Centreplate Model approach seeks to identify where critical points are in food supply chains with respect to commodity resource flows and it has been used to assess national food systems (Martindale *et al.*, 2020c). This has proved useful because it is possible to investigate food categories at national scale that have limited dietary relevance, and Centreplate Model approaches provide a viewpoint that enables a focus on food categories that are most critically controlled in supply chains with respect to a specific nutrient group and supply attribute, e.g. protein supply and production (Table 3.1).

Table 3.1. The rank of protein supply for the global food system across different food material categories, the rank score of protein supply (g. capita^{-1}.day^{-1}) benchmarked against production and food-loss rank scores. The data was obtained from FAOStat for 2013 categories. The rice category is for milled equivalent mass, milk mass excludes butter, and groundnuts are shelled equivalent mass. For the global production, ranked over 96 is zero and food loss ranked over 78 is zero, and therefore not applicable (n/a). The rank values have no units because they are ratios. The light grey shading of the production data shows points where production rank is lower than protein rank; dark grey shading shows where production rank is greater than protein rank. (Adapted from Martindale *et al.*, 2020c)

	Rank value from FAOStat 2013		
Global rank	Protein supply g/day/capita	Production*	Food loss
1. Wheat and products	15.87	5	6
2. Rice	10.13	6	7
3. Milk	8.22	4	8
4. Poultry	5.16	17	46
5. Pork	4.65	16	49
6. Vegetables, other	3.96	3	1
7. Maize and products	3.58	2	3
8. Beed	3.54	26	43
9. Eggs	2.79	24	20
10. Pulses, other and products	2.25	35	27
11. Freshwater fish	2.04	31	NA
12. Potatoes and products	1.5	7	4
13. Beans	1.47	50	32
14. Soybeans	1.35	8	18
15. Pelagic fish	1.14	36	n/a
16. Groundnuts	1.1	37	30
17. Offal, edible	1.1	55	60

3.4. How a Farm-to-taste Trusted Assessment Has Developed the Digital Twin Methods to Provide Value-added Resource-flow Insights

It can be demonstrated that farm-to-fork approaches align with GIS-LCA methods and connecting these to consumer responses is a future goal of farm-to-taste (Martindale 2017a). This needs to be system-wide, and feedback regarding the taste, aroma, nutrition and dietary impacts – which are the consumer's experience – could be reported using digital technologies that already collect retail sales data efficiently. This has become important because of GHG emissions associated Scope 3 or supply chain emissions. Distributed ledger technology (DLT) and blockchains have demonstrated the capability to provide both traceable and tracking functions for high-volume and high-variability financial flows in systems and report them as immutable data. It was thought that Scope 3 emissions were far too variable and complex to deal with but with these technologies they are not, so their impact is significant in understanding how to project resource flows in methodologies such as Digital Twins (Martindale *et al.*, 2020a).

In the future, LCA application in policy development will map the spatial impact of consumption in urban environments. Martindale *et al.* (2020a) have reported the mapping of food consumption in urban areas using geographic information systems (GIS), LCA outputs and census data. This development brings us the potential to formulate digital twins for consumption so that organizations can forecast production and consumption with confidence, which is of great use because they will remove wasteful practices from the food system, i.e. enable supply chains so that supply meets demand precisely. This has been considered far too complex before Industry 4.0, and improved access to IDTs has also been limited by the notion that the food system is not clustered. Geo-spatial analysis of the food industry and populations shows that it is very much clustered and not randomly developed (see Fig. 3.3, for the poultry industry, one of our top three proteins supplied in the UK). Food is produced and manufactured in distinct regions because of access to resources and this results in valleys and clusters. Methods to assess how these are connected have been developed so that they can be used to assess the efficiency of the food system (Martindale *et al.*, 2020a). Understanding how clusters of activity are connected is crucial if we are to deliver sustainability or a circular economy, both of which depend on the efficient connectivity of different resources so that transfer of materials is achieved in the most energy-efficient way. There are numerous examples of such clusters and their importance has been established for co-design of food and feed systems in a circular economy utilizing field vegetable production for fresh produce production and insect larvae production (Jagtap *et al.*, 2021).

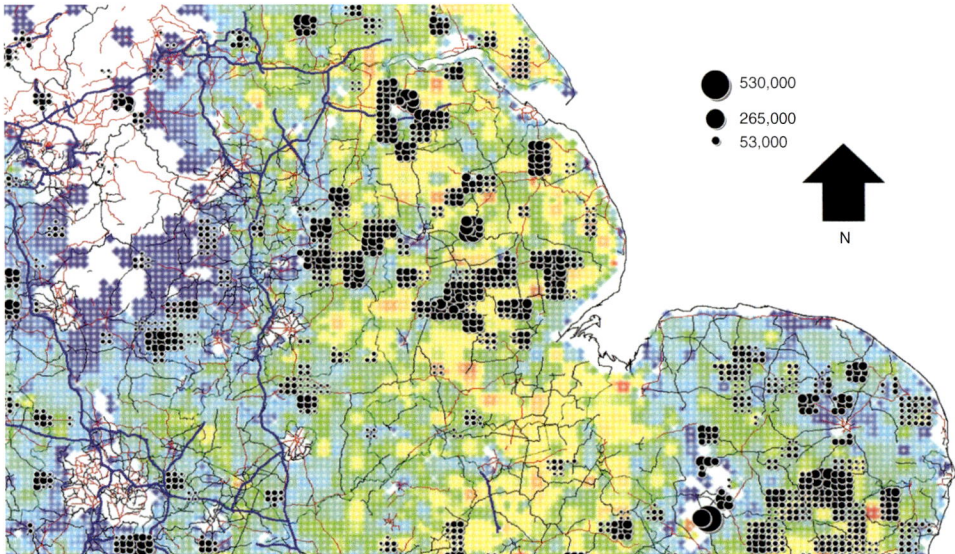

Fig. 3.3. Geo-spatial analysis of wheat production where the coloured 2 km^2 grid shows wheat intensity at blue (lowest) 0 ha, to red (highest) at 510 ha per 2 km^2; and poultry production in the east of England (UK). The black circles indicate the intensity of poultry production for each 2 km^2 with the range of 53,000–530,000 table birds per 2 km^2.

Not knowing where resources are, where they are going and how they are being used means there will be little sustainability associated with outcomes.

3.5. Building a Sustainability Index for the Perfect Meal

Building in sustainability during product development requires guidance. There are established indices of food sustainability that operate at national scales that need to be embedded or re-scaled into business operations for this to happen. Without engaging all food businesses, a sustainable food system will not be possible. Hence, the measurement of sustainability needs to be embedded in the culture of an industry, and a sustainability index scorecard approach can help to achieve this. Testing the index shown here has identified attributes that can be assessed by collecting quantitative and qualitative data by food and beverage manufacturers. The universal use of such a measure would be necessary because the use of consumer goods is interwoven into all of the 17 SDGs, and while consumer goods improve the lives of billions of people, an inevitable outcome of their manufacture is increased consumption and utilization of natural resources (Martindale, 2017a). For many consumers, this is perceived as a net depletion of resources despite a legacy of sustainable reporting that demonstrates that a circular economy is possible. Net-zero and carbon-neutral manufacturing is now possible but reducing consumption when much of the world aspires to increase it remains antagonistic. Food and beverage products provide specific issues where global indices of nutrition and food sustainability have already facilitated practical actions that improve food security. The methodologies to do this are applied at a national scale, and the methods demonstrated here can be used from product concept to consumer to achieve sustainable consumption goals. The idea of assessing the sustainability scope of product development ambitions within minutes is required by companies taking directional decisions daily and production operators making operational choices at each production period. This must be achieved expediently, and the ability to guide these decisions for sustainability attributes and consumer experience are essential to every food and beverage business. The current national index measurements have been developed so that their scope can be applied to food categories and be used by food processors and manufacturers for specific food supply chains. The focus is to apply these to decisions made across the food product development and manufacturing arena so that greater sustainability is an outcome. Current indices of food security and sustainability not only identify high-level policy risks but also focus on the production of agricultural commodities and do not provide detailed insight into the resilience of the manufacturing, distribution and retailing functions of supply chains. This means that such static or state indicators remain of use as a snapshot of the food system, but the potential to collect data from the system all the time means there is a need for indicators that can assess food security, nutritional goals and sustainability at the product development level.

The established indices include the Global Food Security Index (GFSI), the global Access to Nutrition Index (ATNI) and the Food Sustainability Index (Gustafson *et al.*, 2016; Chaudhary *et al.*, 2018; Haddad, 2018; Chen *et al.*, 2019). These have provided the platform in the sustainability index shown here so that the impact of these high-level indices is extended to individual manufacturers and product developers. In the UK, there are over 350,000 food businesses; the goal of the sustainability index is to improve security and sustainability in each of these food businesses when products are at concept, development or manufacturing stage. The application at such an operational level must be open-sourced and available to food businesses because there are known barriers where current indices for nutrition, security and sustainability exist, but they are only typically used by large food groups and companies. In an open-data food system, access to methods for assessing food products' sustainability must become available across the food system, and this is often limited by resourcing and skills. Another method for bringing together several attributes of sustainability in the food system is to consider the ecosystem services associated with products. These are well aligned to food and beverage since they are derived from land-use change and land utilization where these have been considered part of the agricultural system since 1926, when Transeau first produced his treatise on embodied energy of maize crops (Transeau, 1926). This systemic approach has been tested and refined leading to highly accurate determinations of farming energy balances and carbon footprint of European agriculture (Brentrupp *et al.*, 2004). LCA methodologies are providing an option for policymakers who require a measurement of efficiency (Clune *et al.*, 2017), but a real game-changer has been a structured framework for assessing ecosystem services associated with food supply chains and product life-cycles (Costanza *et al.*, 1997).

Ultimately, supply chain metrics need to be aligned with consumers' requirements and this needs methodologies that can assess and measure the experience of food and beverages. Sensory testing associated with product development can be in-depth and complex with established studies qualifying the value of consumer panels and trained assessment panels. The alignment of this type of assessment with sustainability outcomes requires a simplified approach to capture data on preference and utilization of foods. There are established studies that identify flavour or odour attributes of foods in different food categories and these can be used to guide integration of sensory with sustainability outcomes (Aparicio and Morales, 1995; Lawless *et al.*, 2012; Mojet and de Jong, 1994; Nieva-Echevarría *et al.*, 2017). This approach can be extended to baked and roasted products which provides innovative approaches to food insights because it can be aligned with the energy used in cooking and processing specific foods and beverages (Ladha-Sabur *et al.*, 2019). An innovative approach to obtaining data from consumer panels is to use 'check all that apply' structures, which ask consumers to indicate what preferences relate to specific flavours, colours and textures. These types of survey can also provide context in terms of amount of food served on a plate or service; for example, is this less than 0.75, 0.5 or 0.25 of the portion expected? This type of survey has been used successfully to assess the amount of food wasted for each serving in domestic settings (Martindale, 2014; Martindale and Schiebel, 2017). An extension to this approach is to really integrate fulfilment values into consumer surveys using emotional cues associated with foods and beverages. This is well developed for specific brand owners who understand what quality and fulfilment cues consumers most respond to for specific products. The approach has also been developed in research with descriptive emotional codes derived from King and Meiselman (2010). An example of this coding is shown in Table 3.2.

The development of a Quality Index is a goal for many food and beverage organizations who are looking to embed sustainability into workforce activities, manufacturing processes and places of work. A straightforward approach to developing these types of indices is demonstrated here and it is compared to the six-function model (6FM) presented by Martindale (2017a) and used to assess energy embodied in food and beverage products manufactured. The 6FM considered sourcing ingredients, manufacture, distribution, retailing and utilization (domestic cooking and wastes). It has aligned these LCA-focused metrics with nutrition and food choices made in diets. The six functions within the model accommodate this complexity and it has been tested with manufacturers and retailers. The model was unnecessarily complex because decisions made in manufacturing or retailing environments are made expediently and require guidance on bringing relevant data sources into the product-focused processes. This is what the Quality Index shown here attempts to achieve by focusing on analytical components of sustainability metrics while integrating these with an assessment of ethical values associated with food and beverage products. The model for the Quality Index uses the following terms to summarize the attributes used to calibrate quantitative and qualitative data where Q equals quality, A equals analysis and E equals ethics.

$$Q = A + E$$

They are described as follows. **Quality** is the outcome of the quality system and focus in products and processes. It is recognized that the relationship between quality, analysis and ethics must be simplistic so that they can be used quickly and efficiently. The **analysis** function is composed of three calibrated attributes that rely on obtaining food system data from external databases for nutrient content of product, trade flows associated with the main category needed for the product, GHG emissions associated with the product, and the risk that the product will be wasted. The three calibration attributes are as follows:

Table 3.2. An example of selecting emotional coding associated with food and beverage products. (Adapted from King and Meiselman, 2010)

Set 1	Set 2	Set 3	Set 4	Set 5	Set 6
happy joyful merry	pleased friendly glad	peaceful warm calm	satisfied mild steady	adventurous energetic enthusiastic	active aggressive eager

1. Nutritional density (ND) and GHG intensity: health and well-being is still a major 'trend' in consumption choice because there is, typically, a good understanding that it is based principally on energy density or calorie content. Using a nutrient density model (ND15) refines this understanding so that we can quantify and calibrate broader measures of nutritional value (Drewnowski et al., 2015).

2. Import and export intensity (IE): these data are obtained from national statistics and they are reported as a percentage of imports with respect to domestic stock quantity (production + imports + stocks – exports). This has been tested in the Centreplate Model where import and export intensity are ranked with respect to consumption demand (Martindale et al., 2020a).

3. Waste risk (WR): these data use the waste ranking for specific food categories where the data is difficult to define and reported variably by different sources (Martindale, 2017a).

Lastly, ethics is assessed using a matrix of certification and standards utilized in the food and beverage industry which is currently being developed but utilizes established certification such as Rainforest Alliance, Marine Stewardship Council, UTZ and numerous fair trade or resource-management protocols (an example is shown in Table 1.1). The Q = A + E systematic approach can guide product design and support existing certifications and open-source aspects of commercial certification. The development of different channel segments such as frozen, chilled, ambient and fresh are being tested using Q = A + E and the 6FM.

3.6. The Requirement for a Balanced Global Diet that Connects 9 Billion Consumers

Best practice in the food and beverage industry has been transformed by sustainability. It resonates across industry and consumers as an ideal that we should rightly strive to achieve. Much of what we have been aiming for is to reduce the GHG associated with the production and consumption of foods. Manufacturers are now reporting carbon-zero product categories, including whole milk and beef, which was unthinkable ten years ago. Our improved understanding of how resources flow through food systems has made carbon-zero a reality.[1] Programmes that sought to reduce GHG emissions ten years ago exposed many gaps in our understanding of food systems. The initial debates tended to demonize food and beverage products with increased carbon footprints, namely livestock products and beef. What these studies do not consider is nutritional delivery and consumer experience, both of which are important because without them sustainability will never be delivered. This is because every meal must provide balanced nutrition and a favourable experience. If it does these two things, it is more likely that it will not be wasted and result in optimal health. Carbon-zero thinking has been transformative in breaking the deadlock between carbon footprint and nutrition with the launch of branded zero-carbon livestock products such as whole milk, beef and lamb. These have shown that food producers and manufacturers are confident in claiming it.[2]

The subsequent rethinking of carbon footprinting is enlightening because it can be related to achievable and nutritious diets and lifestyles so that responsible consumption is possible. Improvements must not get lost in purely carbon-footprinting diets. We are developing models for the UK that identify where critical points and connectivity in the food system control resource flows (Martindale et al., 2020a). New product development (NPD) is the operational activity we are focusing on because if product developers and technologists build in sustainability at the concept stage, there is an increased possibility that the final product will deliver it (Jagtap and Duong, 2019). One of our models – Centreplate – is currently being tested to NPD strategies, improving protein supply and reducing waste (Martindale et al., 2020b). We are now at a point where food system insights have the potential to bring sustainability and nutritional datasets together because of two technological advances we would consider most notable. The first is the ability to embed digital technologies into resource packaging so that traceability and analysis of supply chain data can be enabled securely for most food companies (Martindale et al., 2018). The other is the projection of the dietary impact of nutrition on populations where product development has a controlling role (Jagtap and Duong, 2019). This changed forever a generation ago in response to the newly sequenced human genome and what followed was a scramble for therapeutics, but the interaction of health and nutrition through our diet was largely overlooked by all of this (King et al., 2017). We now have a greater understanding of how genes and metabolism interact with what we choose to eat. It is essential to keep the food system lens, which the Sustainable Nutrition Initiative's Delta Model does (Smith et al., 2020). Connecting datasets and making sure we speak to each other is becoming increasingly important. This is otherwise known as interoperability in the digital arenas. We can deliver a net-zero sustainable food system, but without interoperability it will not happen; our food future depends on all partners in the global network connecting methods and data that will guide sustainable dietary choices.

3.7. Conclusion

The requirement to define and measure resource utilization in food and beverage supply chains has a legacy of very clear commercial motivation to apply methods that improve optimization and efficiency improvements. The connection between resource flow and product occurs in product development and if these principles are built into product development, then sustainable outcomes are more robust, with reduced waste, improved nutrition and reduced carbon footprint of consumption. There has always been a different perspective in food and drink manufacturing than in other consumer goods because access to them is a human right and product quality has such a large impact on consumer experience. The influence of shelf-life, product stability and packaging were exceptionally important among all of the usual drivers of preference that include the dominance of branding. This has traditionally made it difficult to measure the impact of resource flows in food and beverage supply chains beyond the commercial goals of financial optimization. This worldview changed completely with the development of ecosystem services and green capital assessments by the 'green giants' of the business ecosystem (Williams, 2015). The importance of balancing production and consumption for sustainable outcomes became a values mission for many companies and this has since become one that reduces waste in all its forms. The methods introduced to assess this new approach provided added values to resource flows where there was a growing realization that any solutions to optimizing resource use in a globalized economy means a system-wide view is required. It has since been demonstrated that added value is obtained through collaboration and co-operation of not only materials or commodities but also innovation, finance, people and business resources. These are all well characterized processes; they have been so for over 50 years and are what should be known as landmarks in our development of Digital Twins (May, 2019).

When theoretical ecologists first used computers to scale up simple mathematical relationships between growth and resources to population level they began the route to developing the Digital Twins demonstrated in this chapter. These can guide sustainable outcomes through the chaotic and complex outcomes of scaling up of meta-NPD in food systems. The issue in getting to this system viewpoint was always the methodologies used, which the food system had now developed with LCA, and the instruments to deliver them to every food and beverage company, which the system did not have. That was until now. This chapter has reported and demonstrated the digital tools that can assess and construct guidance from large datasets of resource flow in our global food system. The use of geographical information, Digital Twin approaches and assessment methodologies such as LCA have been tested for food and beverage resource flows here. What is changing in our food and beverage industry is the ability to package these tools into digital applications that can be accessed by millions of end-users in a secure way to enhance the credibility and trust associated with data. This has developed the blockchain and other technologies that were initially tested in finance sectors where trust was critical – in digitized monetary transfers, for example. They have become an application that is being used by food and beverage consumer goods companies which is transformative because it enables the transfer of resource-flow information that may be relevant to specific stakeholders in a food or beverage supply chain, e.g. carbon footprint information. The impact of these technologies is yet to be shown because even though the technologies offer an instantaneous assessment of food system inventory or an instantaneous audit of supply chain operations, they are yet to become mainstream applications. The technologies that instantaneously assess how much resource is where in a global system are of high value to commercial operators and the high-level policy goals such as the UN SDGs for example. They offer so much promise and, as ever with any resource flow or supply chain focus event, they will depend on collaboration and co-operation.

The technology is in place but the frameworks of equitable collective, from smallholder groups to global groups, are yet to be demonstrated. As ever, demonstration is critical to acceptance and the next few years will determine how such collectives in food and beverage can be made for worldwide resource use that delivers sustainable production and consumption.

Notes

[1] New Zealand's first carbon-zero milk. See: https://www.fonterra.com/nz/en/our-stories/articles/new-zealands-first-carbonzero-milk.html (accessed 18 December 2020).
[2] CN30 overview. See: https://www.mla.com.au/research-and-development/Environment-sustainability/carbon-neutral-2030-rd/cn30/ (accessed 18 December 2020).

References

Alghamdi, D.A., Dooley, D.M., Gosal, G., Griffiths, E.J., Brinkman, F.S.L. and Hsiao, W.W.L. (2017) FoodOn: A Semantic Ontology Approach for Mapping Foodborne Disease Metadata. Available at: chrome-extension://oemmndcbldboiebfnladdacbdfmadadm/http://ceur-ws.org/Vol-2137/paper_32.pdf (accessed 23 May 2022).

Aparicio, R. and Morales, M.T. (1995) Sensory wheels: a statistical technique for comparing QDA panels: application to virgin olive oil. *Journal of the Science of Food and Agriculture* 67, 247–257.

Bouis, H.E., Saltzman, A. and Birol, E. (2019) Improving nutrition through biofortification. *Agriculture for Improved Nutrition: Seizing the Momentum* 47.

Brentrup, F., Küsters, J., Kuhlmann, H. and Lammel, J. (2004) Environmental impact assessment of agricultural production systems using the life cycle assessment methodology: I. Theoretical concept of an LCA method tailored to crop production. *European Journal of Agronomy* 20, 247–264.

Chaudhary, A., Gustafson, D. and Mathys, A. (2018) Multi-indicator sustainability assessment of global food systems. *Nature Communications* 9, Art. 848.

Chen, P.-C., Yu, M.-M., Shih, J.-C., Chang, C.-C. and Hsu, S.-H. (2019) A reassessment of the Global Food Security Index by using a hierarchical data envelopment analysis approach. *European Journal of Operational Research* 272, pp. 687–698.

Clune, S., Crossin, E. and Verghese, K. (2017) Systematic review of greenhouse gas emissions for different fresh food categories. *Journal of Cleaner Production* 140, 766–783.

Costanza, R., d'Arge, R., De Groot, R., Farber, S., Grasso, M. *et al.* (1997) The value of the world's ecosystem services and natural capital. *Nature* 387, 253–260.

Dooley, D.M., Griffiths, E.J., Gosal, G.S., Buttigieg, P.L., Hoehndorf, R. *et al.* (2018) FoodOn: a harmonized food ontology to increase global food traceability, quality control and data integration. *npj Science of Food* 2(1), 1–10.

Drewnowski, A., Rehm, C.D., Martin, A., Verger, E.O., Voinnesson, M. and Imbert, P. (2015) Energy and nutrient density of foods in relation to their carbon footprint. *American Journal of Clinical Nutrition* 101(1). Available at: https://doi.org/10.3945/ajcn.114.092486 (accessed 23 May 2022).

Espinoza-Orias, N. and Azapagic, A. (2017a) Understanding the impact on climate change of convenience food: carbon footprint of sandwiches. *Sustainable Production and Consumption.*

Espinoza-Orias, N. and Azapagic, A. (2017b), "Understanding the impact on climate change of convenience food: Carbon footprint of sandwiches", *Sustainable Production and Consumption* 15, 1–15.

Finglas, P.M., Berry, R. and Astley, S. (2014) Assessing and improving the quality of food composition databases for nutrition and health applications in Europe: the contribution of EuroFIR. *Advances in Nutrition* 5, 608S-614S.

Garnett, T. (2016) Plating up solutions. *Science* 353, pp. 1202–1204.

Griffiths, E.J., Dooley, D.M., Buttigieg, P.L., Hoehndorf, R., Brinkman, F.S.L. and Hsiao, W.W.L. (2016) FoodOn: a global farm-to-fork food ontology. *ICBO/BioCreative*, pp. 1–2.

Gustafson, D., Gutman, A., Leet, W., Drewnowski, A., Fanzo, J. and Ingram, J. (2016) Seven food system metrics of sustainable nutrition security. *Sustainability* 8, 196.

Haddad, L. (2018) Reward food companies for improving nutrition. *Nature* 556, 19–22.

Jagtap, S. and Duong, L.N.K. (2019) Improving the new product development using big data: a case study of a food company. *British Food Journal* 121, 11.

Jagtap, S., Garcia-Garcia, G., Duong, L., Swainson, M. and Martindale, W. (2021) Codesign of food system and circular economy approaches for the development of livestock feeds from insect larvae. *Foods* 10(8), p. 1701.

King, S.C. and Meiselman, H.L. (2010) Development of a method to measure consumer emotions associated with foods. *Food Quality and Preference* 21, 168–177.

King, T., Cole, M., Farber, J.M., Eisenbrand, G., Zabaras, D., Fox, E.M. and Hill, J.P. (2017) Food safety for food security: relationship between global megatrends and developments in food safety. *Trends in Food Science & Technology* 68, 160–175.

Kinouchi, O., Diez-Garcia, R.W., Holanda, A.J., Zambianchi, P. and Roque, A.C. (2008) The non-equilibrium nature of culinary evolution. *New Journal of Physics* 10(7), p. 73020.

Krzywinski, M.I., Schein, J.E., Birol, I., Connors, J., Gascoyne, R. *et al.* (2009) Circos: an information aesthetic for comparative genomics. *Genome Research* 19, 1639–1645.

Ladha-Sabur, A., Bakalis, S., Fryer, P.J. and Lopez-Quiroga, E. (2019) Mapping energy consumption in food manufacturing. *Trends in Food Science & Technology* 86, 270–280.

Lawless, L.J.R., Hottenstein, A. and Ellingsworth, J. (2012) The McCormick spice wheel: a systematic and visual approach to sensory lexicon development. *Journal of Sensory Studies* 27, 37–47.

Martindale, W. (2014) Using consumer surveys to determine food sustainability. *British Food Journal* 116(7). Available at: https://doi.org/10.1108/BFJ-09-2013-0242 (accessed 23 May 2022).

Martindale, W. (2016) The potential of food preservation to reduce food waste. *Proceedings of the Nutrition Society* 76, 28-33. Available at: https://doi.org/10.1017/S0029665116000604 (accessed 23 May 2022).

Martindale, W. (2017a) Cutting through the challenge of improving the consumer experience of foods by enabling the preparation of sustainable meals and the reduction of food waste, food waste reduction and valorisation: sustainability assessment and policy analysis. Available at: https://doi.org/10.1007/978-3-319-50088-1_2 (accessed 23 May 2022).

Martindale, W. (2017b) The potential of food preservation to reduce food waste. *Proceedings of the Nutrition Society* 76(1), 28–33.

Martindale, W. and Schiebel, W. (2017) The impact of food preservation on food waste. *British Food Journal* 119(12). Available at: https://doi.org/10.1108/BFJ-02-2017-0114 (accessed 23 May 2022).

Martindale, W., Swainson, M., Hollands, T. and Marshall, R. (2018) Bread winner. *Food Science and Technology* 32(3).

Martindale, W., Mark, S. and Hollands, T. (2019) New direction for NPD. *IFST Journal* 33(4).

Martindale, W., Duong, L. and Swainson, M. (2020a) Testing the data platforms required for the 21st century food system using an industry ecosystem approach. *Science of the Total Environment*, p. 137871.

Martindale, W., Swainson, M. and Choudhary, S. (2020b) The impact of resource and nutritional resilience on the global food supply system. *Sustainability* 12(2), p. 751.

Martindale, W., Wright, I., Korir, L., Opiyo, A.M., Karanja, B. *et al.* (2020c) Framing food security and food loss statistics for incisive supply chain improvement and knowledge transfer between Kenyan, Indian and United Kingdom food manufacturers. *Emerald Open Research* 2(12), 12.

May, R.M. (2019) *Stability and Complexity in Model Ecosystems*, Vol. 1. Princeton University Press.

Mojet, J. and de Jong, S. (1994) The sensory wheel of virgin olive oil. *Grasas y Aceites* 45(1).

Nieva-Echevarría, B., Manzanos, M.J., Goicoechea, E. and Guillén, M.D. (2017) Changes provoked by boiling, steaming and sous-vide cooking in the lipid and volatile profile of European sea bass. *Food Research International* 99, 630–640.

Price, S. and WHO (2020) Codex and the SDGs: how participation in Codex Alimentarius supports the 2030 Agenda for Sustainable Development. World Health Organization.

Rejeb, A., Keogh, J.G., Martindale, W., Dooley, D., Smart, E. *et al.* (2022) Charting past, present, and future research in the semantic web and interoperability. *Future Internet*. Available at: https://doi.org/10.3390/fi14060161 (accessed 4 June 2022).

Rivera, X.C.S. and Azapagic, A. (2019) Life cycle environmental impacts of ready-made meals considering different cuisines and recipes. *Science of the Total Environment* 660, 1168–1181.

Smith, N., Fletcher, A., Finer, O. and McNabb, W. (2020) Sustainable nutrition initiative-feed our future. *Food New Zealand* 20(5).

Song, A., Xue, G., Cui, P., Fan, F., Liu, H. *et al.* (2016) The role of silicon in enhancing resistance to bacterial blight of hydroponic-and soil-cultured rice. *Scientific Reports* 6, p. 24640.

Transeau, E.N. (1926) It has therefore seemed profitable to study this problem of the accumulation of energy by plants both for what information. *The Ohio Journal of Science* 26.

Tuwani, R., Sahoo, N., Singh, N. and Bagler, G. (2019) Computational models for the evolution of world cuisines. 2019 IEEE 35th International Conference on Data Engineering Workshops (ICDEW). IEEE, pp. 85–90.

Williams, E.F. (2015) *Green Giants: How Smart Companies Turn Sustainability into Billion-Dollar Businesses*. AMACOM Division of American Management Association.

4 Food 4.0: Industry 4.0 Applications in the Food Sector

Sandeep Jagtap

4.1. The Fourth Industrial Revolution: What Does It Mean?

The fourth industrial revolution, also termed Industry 4.0, is transforming the way we produce goods through digitization, which makes it possible to collect and analyse data across machines, sensors and IT systems. It allows faster, flexible and efficient operations to manufacture high-grade products at reduced costs. Like other manufacturing sectors, the food sector is embracing Industry 4.0, i.e. Food 4.0, to increase productivity, growth, shift and workforce efficiency in order to remain competitive in the fast-changing market dynamics. This chapter presents the nine technologies that form the building blocks of Food 4.0, i.e. autonomous robots, simulation, horizontal and vertical integration, Internet of Things (IoT), cybersecurity, the cloud, additive manufacturing, augmented reality, and big data and analytics. Furthermore, this chapter will highlight the application of the mentioned nine technologies in the food sector. However, the solutions will vary depending on the factory set-up and equipment installed. Some of the examples are food safety, traceability and transparency, warehouse automation, preventive maintenance, tracking goods and inventory, intelligent manufacturing and other functions.

4.2. Overview of Industry 4.0

In this recent era of globalization, the food sector is under constant strain to advance, improve competitiveness and perform better than rivals in the worldwide market. Industry 4.0 brings an important set of applications in order to achieve these endeavours. These can assist food sector through robotics and automation, digitization, reduced errors in processes, an improved ability to make an informed decision, improved efficiency and productivity, decreased costs and in general enabling more to be accomplished with less (Sharma and

© CAB International 2022. *Food Industry 4.0: Unlocking Advancement Opportunities in the Food Manufacturing Sector* (W. Martindale et al.)
DOI: 10.1079/9781789248593.0004

Jain, 2020). Furthermore, Industry 4.0 has resulted in innovative processes, products and services. It has transformed the industry, leading to improved productivity. This is what has motivated several large food companies to follow the digitization approach as a means to remain competitive and sustainable (Jagtap *et al.*, 2021a).

Industry 4.0 is one of the big buzz words that is flying around at the moment and it basically refers to the fourth industrial revolution, where we have seen transformative technologies that have changed industries. This fourth industrial revolution is all about the smart factories with connected machines and intelligent robots. As shown in Fig. 4.1, it all started with the first industrial revolution in the 1770s, where we moved from working by hand to machines, from farms to factories, using steam and hydropower. Then, 100 years later, electricity arrived in the next industrial revolution, which gave us automation and assembly lines, and this was the start of the factories. Another 100 years later, computers and electronics arrived around the 1970s, which allowed automation of some of the blue-collar work in the factories. And the latest industrial revolution, which is referred to as Industry 4.0, is not looking at the individual computerized machines but at the whole network of them. In short, all the machines are talking to each other which is termed 'intelligent factories'. Some of the underlying technologies of this industrial revolution are IoT and Big Data, and others will be mentioned in following sections.

The fourth industrial revolution was triggered by the development of Information and Communications Technologies (ICT). Cyber-physical systems (CPS) ensured smart automation with decentralized control and advanced connectivity (IoT) formed the underlying technological basis. It resulted in

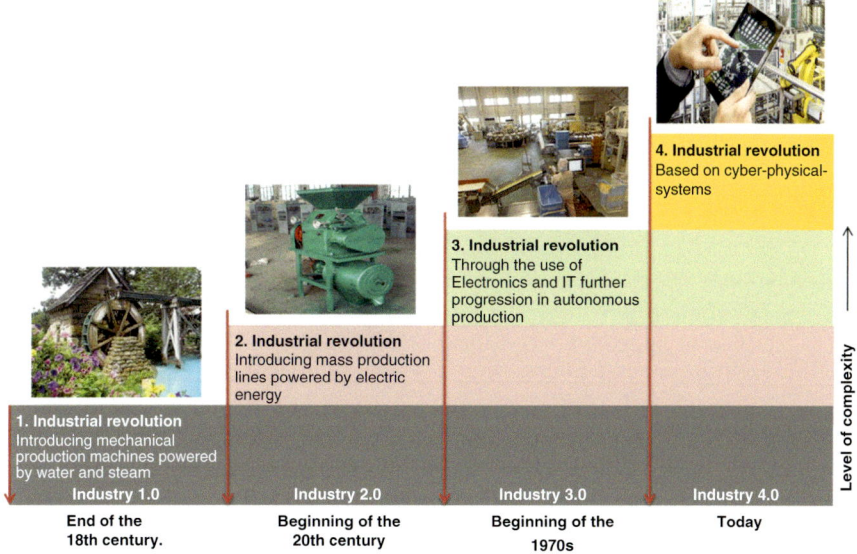

Fig. 4.1. Industry 4.0 Application. (Jagtap, 2021; created for this book)

reorganization of traditional automation systems to self-organizing CPS-based production systems that permit flexible and efficient production of goods.

The Industry 4.0 concept originated in Germany, which is one of the leading nations adopting this concept. Germany has consolidated its reputation as the world leader in the manufacturing sector and the world's factory outfitter; it is a nation equipped to meet the challenges of the digitization era. In 2011, the Industry 4.0 concept was first coined at the Hannover fair (Zhou et al., 2015). Since, then it has become an area of interest for researchers, academics and industrialists all over the world. The fundamental idea is to utilize the potentials of new technologies; for example, accessibility and utilization of the IoT, integration of production and operational activities, digitization and transparency, and efficient production with quality products. Although Industry 4.0 has resulted in digitization and novel technologies with a primary focus on generating profits, there is still potential to explore new opportunities. Adoption of the just-in-time approach and other lean management techniques combined with outsourcing production to countries with lower production costs have reduced production costs. But adoption of Industry 4.0 can also solve this problem and in some cases has resulted in reduction of production, logistics and quality management costs by between 10% and 30%. There are other benefits of adopting the Industry 4.0 approach, such as shortest possible time to launch a new product in the market, better customer responsiveness, mass production with lower production costs, and flexible, efficient and sustainable production (Rojko, 2017).

Industry 4.0 stresses the concept of digitization and linking of all productive units in a factory or an economy. As shown in Fig. 4.2, there are several technological areas that underpin Industry 4.0, which are robotics, big data analytics, additive manufacturing, IoT, cybersecurity, augmented reality, cloud computing, horizontal and vertical system integration, and simulation.

4.3. Robotics and Automation in Food

Advances in the food manufacturing sector during the last few decades have prompted the utilization and implementation of robotization and automation. The incessant need to improve sustainability and efficiency within the food sector has prompted the implementation of robots and automation within the supply chain. According to the most recent figures published by the British Automation and Robotics Association (BARA), in 2013 there was a 60% increase in adoption of robotics by UK food manufacturers compared to the year 2000 (Industrial Technology, n.d.). Within food manufacturing, the use of robots is mostly at the end of the production line, i.e. for packing and/or palletization of the finished food product. Kuka, a German automation company, developed a robot named KR 3 Agilus, which it claimed to be the first fast-food robot machine to serve food, drinks and desserts to guests (*Industry Europe*, 2020). KR 3 Agilus is a robot cook developed for Bionicook and can perform a number of activities, such as serving food and beverages to customers, at a very low cost. In the UK food industry, robots are limited to packing

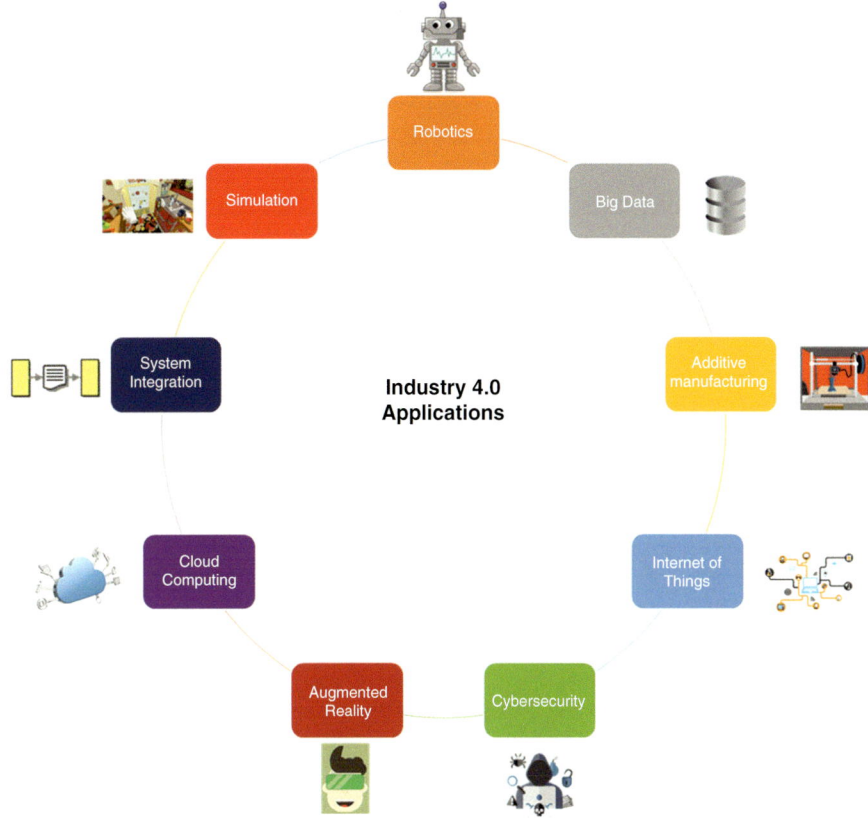

Fig. 4.2. Industry 4.0 Application. (Jagtap, 2021; created for this book)

of finished products for dairy, beverages, confectionery and canned foods. The Flex Picker robot, launched in the year 1998, was considered revolutionizing, as it was the quickest pick-and-place robot (Brantmark and Hemmingson, 2001). Potential advantages of implementing robots and automation are in operational efficiency by reducing staff, material and vehicle movements as well as reducing food processing activities.

Nowadays, food manufacturers use cost effective but simplistic automation solutions to generate higher outputs when compared to traditional food factories (Duong *et al.*, 2020). Due to Brexit and the COVID-19 pandemic, there is a shortage of food manufacturing workers in the UK, who are considered a traditional labour force in food production (Trollman *et al.*, 2021). Therefore, robotization and automation are the preferred options for food manufacturers. Other than pick, place, packing and palletization, robots and automation are being used for food processing operations such as cutting, chopping, filling, cooking, measuring, etc. (Buckenhüskes and Oppenhäuser, 2014).

Agri-robots are currently being used on farms for seeding, watering, harvesting, cutting, chopping, processing and packing of food products (Kondo *et al.*, 2011). Figure 4.3 shows various kinds of robots being used in the bakery

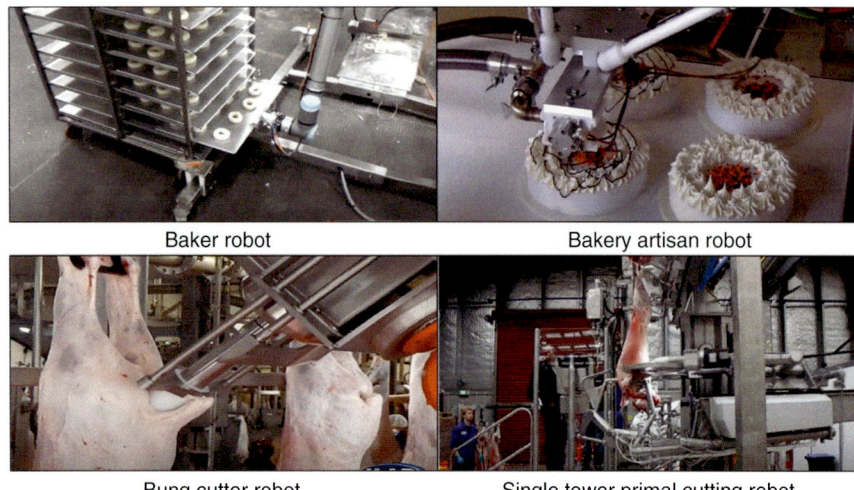

Fig. 4.3. Bakery processing robots. (Jagtap, 2021; created for this book)

processing industry to perform activities such as placing products onto a baking trolley, decoration, sauce spreading, mixing and cooking. Similarly, in the beverage industry robots are used for cleaning, filling, labelling and packing the containers into boxes (Saravacos and Kostaropoulos, 2016). Modern camera vision systems are used for identifying defects and quality inspection. Saldaña *et al.* (2013) discusses vision systems that are used for grading, inspecting and controlling the quality of vegetables and fruits. As shown by Gebbers and Adamchuk (2010), food manufacturers have increased their productivity by 25% through deployment of robotics and automation rather than using human labour; however, implementation differs in various food sectors. Several factors such as scale of automation, number of robots installed and frequent product variation due to customer needs and demands may affect productivity. For instance, an Argentinian pasta manufacturer enhanced its productivity by 10% by deploying six robots (ABB, 2014). However, most of the food manufacturers need to accommodate a wide product range without altering the production line or set-ups. The current situation, encompassing Brexit, COVID-19, market share and competitive challenges, shows that the food sector should actively pursue investment in robotic automation. Adopting robotics and automation can secure the future of food businesses and contribute positively to the environmental sustainability of food supply chain.

4.4. Big Data in Food

Big Data in simpler terms means the discipline of gathering large volumes of data and transforming it into smaller but meaningful and manageable pieces of information, which can be utilized to achieve detail and appropriate deeper insights on a topic. Big data analytics power food manufacturers by allowing them to make better decisions with regards to pricing, promotion, product development, planning and forecasting. Many of the big data analytics benefits comprise product innovation; increase in sales, marketing, margins and profits; and better customer reach, satisfaction and loyalty (Jagtap and Duong, 2019).

In the food business, timing is crucial, i.e. getting the food product transported to the customer on time is very important due to the limited shelf-life and deteriorating nature of food products. Regardless of all the logistics involved in food production, food hauliers now have access to all the advanced technologies that can help them to improve their efficiency. In order to support food businesses with fast deliveries, big data analytics can be deployed. Some of the elements could be of help in understanding and monitoring of road traffic, weather conditions, route diversions, climate and travelling distance. Considering all the mentioned factors and the information available, an estimated time of delivery can be determined using Artificial Intelligence-based systems. FreshDirect, a company based in USA, deploys sensors to obtain crucial data, which it processes and analyses for monitoring product and environmental conditions during transportation (Splunk, n.d.). Blue Apron, a meal-kit delivery service, uses big data analytics to make real-time decisions about food deliveries, which has resulted in saving a day due to faster decision-making (Google Cloud, 2016).

Tools are employed by food businesses for collecting and understanding information with regard to their customers' choices and opinion of their food products and brands, such as Natural Language Toolkit (Tao *et al.*, 2020). It helps in interpreting customer responses and feedback of their products on social media networks like Twitter, Facebook, etc. It helps in monitoring negative feedback from customers and addressing it immediately before it becomes a bigger problem. This approach is adopted by the majority of fast-food chains, such as Subway, Papa John's, Domino's, Burger King, KFC and McDonald's. For instance, Domino's have their own app, which allows customers to make reservations, place orders and pay for the orders, and offers customers various deals (Samuely, 2017). It allows the company to gain useful customer insights, such as the location visited, what and how often they order, customer experience, time of order and completion of order, and any customer complaints and issues. Quality is considered one of the most important factors within the food industry. It involves storing, processing and handling of food at the right temperature and environmental conditions; any variation in those conditions could lead to damage (Aung and Chang, 2014). IoT-powered sensors can process, analyse and send the data to all the actors in real-time within the supply chain, making it possible to take appropriate actions to prevent any further damage or to replace the damaged products. It can also support in monitoring the quality of incoming and finished food products. In the present scenario, transparency of supply chain actors' activities is necessary for seamless operations. By embedding visibility in supply chains, businesses can boost their product's relationship with their customers and build trust while delivering quality food products (Jagtap, 2019). For instance, Barilla, a pasta and sauce-making company, employed IoT and connected sensors to improve transparency of their supply chain to give consumers greater access to the information about the product and help them identify counterfeit products (Palazzolo, 2015). Big data enables winemakers to preserve and maintain product quality (Hariharan, 2018). For example, it helps in interpreting and predicting the taste and quality of wine, which is reliant on a range of factors such as the right temperature, water, seasons and accurate measurement of ingredients. The insights obtained from the application of big data analytics helps in identifying issues in advance and suggesting preventive measures.

Big data analytics is an important tool for the food sector. It involves all the information with regards to pricing, condition and quality of products, consumer preferences, market trends, brand popularity, etc. (Tirunillai and Tellis, 2014). However, big data requires a thorough analyst and clever software to analyse all the obtained data and make use of it. For example, the predictive analytics powered by the artificial intelligence-based algorithms enables identification of possible problems and results, which gives sufficient time to alter your strategy to obtain desired outcomes (Raza and Khosravi, 2015).

4.5. Additive Manufacturing

The first application of additive food manufacturing was carried out in 2007 by Cornell university researchers. They used a syringe version of an open-source extrusion-based additive manufacturing machine (Periard *et al.*, 2007). Since then, several organizations developed additive manufacturing machines and made available various applications, such as personalized pieces of pasta or chocolate (Pinna *et al.*, 2016). The area of additive manufacturing technology with regards to the food sector is continuously adapting to the designs of food (Godoi *et al.*, 2016). Several commercial solutions are available to the market for both commercial and consumer usage, such as Foodjet by De Grood Innovations, Foodform 3D by RIG, ChefJet and CocoJet by 3D Systems, Foodini by Natural Machines, Choc Creator by Choc Edge, Imagine3Dprinter by Essential Dynamics and Replicator by Makerbot (Lipton *et al.*, 2015; Sun *et al.*, 2015).

The term additive manufacturing – also termed 3D printing – is a process that produces three-dimensional layered physical objects based on computer-aided design (CAD) data (Thymiandis *et al.*, 2012). In 2010, the American Society for Testing and Materials (ASTM) group 'ASTM F42—Additive Manufacturing' articulated a set of standards that categorize the range of additive manufacturing processes into seven categories (ASTM, 2012):

1. Vat photopolymerization
2. Material jetting
3. Binder jetting
4. Material extrusion
5. Powder bed fusion
6. Sheet lamination
7. Directed energy deposition

Additive manufacturing's basic feature is adding material layer-by-layer. However, its corresponding applications can differ and only powder bed fusion (sugar), binder jetting (sugar, protein powders, material extrusion – chocolate, pasta dough) and Vat photopolymerization (egg white, package) are deployed in the food sector. The initial step in the additive manufacturing process is to design a computerized 3D model, which can be transformed into a final product through several stages, without deploying moulds, additional equipment and cutting tools (Piller *et al.*, 2012).

Food printing as 2D printing (laser marking, inkjet printing) has been in use since the 1990s (Flynt, 2020). However, in the last few years the market has seen a rise in additive manufacturing technologies being used in producing a variety of foods, ranging from chocolate to pasta and pizza. The additive manufacturing food system consists of a computer-controlled three-axis motor-powered stage and ingredient feeding mechanism and it operates layer by layer of food as per the design information in its CAD file (Sun *et al.*, 2015). The platform for food printers is based on a Cartesian coordinate system, user interface and layer-forming system consisting of three categories: extrusion, binding and sintering. Material extrusion consists of extruding hot melted food material through a nozzle based on customized geometries, food textures and content (Goyanes *et al.*, 2015). In a hot-melt extrusion process, hot material is pressured through a die of particular design. In binder jetting or inkjet printing food systems, a printhead moves across a bed of food powder (flour) and deposits an aqueous binding food material at desired places. This process is repeated until all layers are finished and unbound powder is removed. Powder bed additive manufacturing systems are used in pharmaceutical applications (Trivedi *et al.*, 2018). Selective laser sintering uses a laser beam to bind materials together to create a solid structure. It is suitable for attaching food components together (Godoi *et al.*, 2016), for example, sintering sugar powder (Pinna *et al.*, 2016).

4.6. Internet of Things

Internet of Things (IoT) is the network of connected devices that collect and transfer data via the Internet (Jagtap, 2019). The food industry is embracing the IoT and several IoT-based applications are available for supply chain actors, including suppliers, manufacturers, warehousing, logistics and retailers, to enhance their operational efficiency and financial performance. For instance, with the aid of Radio Frequency Identification (RFID) tags and transmitters and global positioning systems (GPS), food product movements can be effectively monitored from source to the point of consumption (Zou *et al.*, 2014). It also delivers other advantages, such as knowing customer preferences, market trends and requirements, as well as decreasing surpluses and food wastes. The advanced RFID monitoring provides better visibility of the supply chain, supports automating delivery, recording stocks and despatch activities and real-time tracking and controlling temperature. Through GPS, the exact location of the product can be tracked. By gathering all the useful information, hauliers can predict their operational performance and customer behaviours, and reduce the food miles (Zhou *et al.*, 2013). IoT, therefore, enables the supply chain actors with better visibility of a food product from its origin until it reaches a retailer's shelves. IoT-powered refrigerated containers and warehouses make it possible to monitor and control the temperature of food products remotely (Ramirex-Faz *et al.*, 2020). Any variation in the temperature can be notified immediately in real time, allowing the actors to take necessary action to prevent any further damage and maintain food product quality.

IoT also helps in reducing food waste, which is considered one of the biggest issues in the food sector (Pal and Kant, 2018; Jagtap *et al.*, 2019; Jagtap and Rahimifard, 2019). Approximately one third of the food fit for human consumption is wasted globally. This not only results in financial, energy and water losses, but also contributes to greenhouse gas emissions. IoT makes it possible to monitor the condition of the food products in real time and alerts the respective supply chain actor, who takes immediate actions to reduce waste.

As we are aware, maintenance activities are preventive/or reactive, but not predictive. Therefore, by deploying monitoring sensors we can anticipate issues in advance and address them immediately, saving time and money. IoT solutions in retail such as smart packaging will enable efficient inventory management operations (Biji *et al.*, 2015). It will help in identifying products that are nearing their expiration dates, therefore improving cost effectiveness and efficiency, and reduction in food wastes due to overstocking. Smart packaging also can help to inform the consumer whether a product is nearing its expiration date, indicate if it has gone off, or also suggest a suitable recipe (Kuswandi *et al.*, 2011).

Assigning each food item with their own identity in the form of digital fingerprinting on the Internet not only makes it easy to track, but also makes it possible to detect any food fraud or counterfeit food items. Any supply chain actor will be able to scan digital tags on food packaging to check their authenticity and prevent any fraudulent products from reaching consumers, and consumers will also be able to scan for this information. It helps brands in establishing trust and transparency with their consumers by providing them with authentic and quality food products.

IoT technology is being successfully utilized to monitor temperature and quality of food products. Recently, there has been a lot of development deploying smart sensors in combination with cloud-based analytics to predict pathogenic organisms before a potential outbreak happens. Some food factories, such as SugarCreek Packing Company, installed smart sensors on each piece of equipment and machinery within the factory (Mahadev, 2015). These sensors can identify pathogens from biochemical and chemical reactions during harvesting, processing and distribution stages, thereby ensuring instant removal of pathogens before they become a risk. Usually, purchasers of food products expect transparency from the actors involved in the food supply chain. Therefore, having traceability and transparency in global food supply chains will support food producers to prosper by gaining customer trust and loyalty. Although there are lots of complexities involved in global food supply chains, IoT technology can make it easier for all the actors. Furthermore, IoT brings added benefits, such as better inventory management, cost effectiveness, quicker lead times, identifying and solving flaws in the supply chain, complying with food safety regulations and providing transparency to consumers.

4.7. Cybersecurity

Cybersecurity breaches and website hacking are increasing (Ghadge *et al.*, 2019). The risk from cyber threat and hacking are similar to the food safety risks that supply chain actors mitigate on a daily basis. An approach similar to that used in identifying and addressing food safety risks such as foreign bodies and pathogens can also be applied to shielding the information important for food processing from unscrupulous agents (Jagtap *et al.*, 2021b). These include analysing the cyber threats, understanding the possibility of occurrences and designing a program to alleviate those threats.

The cyber threat could be in the form of data theft, exposure of sensitive data in the public domain, data corruption or loss as well as manipulating and falsifying data. Evaluating the possibility of attack in each case needs recognizing unscrupulous agents and how they may profit from an attack. The unscrupulous agent could be a foreign agent, cybercriminals, unhappy staff, a competitor or fanatical food activist. It is important to understand how these unscrupulous agents can profit from stealing, revealing, manipulating, falsifying or corrupting data.

Unlike other sectors, the threat of data theft is low in food sectors. However, threat to employees personal data stored in financial or HR systems remains high-value, but manufacturing-related data are of low value. Still, companies with special recipes, patented formulas or processes must shield it from falling into the hands of foreign imitators.

The data theft risk may be low in the food sector, but other risks can impact heavily. For instance, the risk of data corruption or loss could result in substantial monetary losses. Failure to show a hazard analysis critical control point (HACCP) check could result in wasting millions of pounds in product recall or product disposal. These monetary risks make company data an area of interest for criminal hackers, who can encrypt the data and hold the data hostage for a ransom. This is also termed a 'ransomware' attack and is an increasing cyber threat. As per *Newsweek*, ransomware attacks increased by over 250% in 2017 compared with 2016 (Cuthbertson, 2017). A ransomware named NotPetya attacked several organizations in the USA, including Merck, a pharmaceutical agent. Merck paid US$ 300 million in 2017 to recover the data (Greenberg, 2018).

Once all the cybersecurity threats are evaluated, a security agenda could be put in place with activities like those similar to ensuring safe and quality food products. Data security experts will perform actions such as sustaining zones of security to avoid threat transfer to other zones, shielding and tracking critical control points and performing regular tests of critical points and the general environment. Similar to the segregation of allergen and non-allergen food raw materials in the factory, the data should also be segregated, and the initial steps are establishing the security zones. For instance, operational information that sits on the same network as staff computers is more prone to risk because virus-infected emails can affect the company's sensitive information. Therefore, segregating sensitive information minimizes access points and handling control points necessitate fortifying the access from data ports to user accounts.

Employing deep security can come at a significant cost if done individually, and requires know-how, hardware, training, continuous monitoring and diligence. Using a cloud-based food quality management system is one of the most economical ways for protecting information rather than relying on on-site systems. Not everyone seems to agree with this tactic; however, cloud-based data management entities are the most well-resourced to protect information. As their business revenues are dependent on protecting their client's data and information, they are heavily engaged in data security. The cloud-based data management entities have an army of data security professionals, data backup systems, frequent system testing, constant system monitoring and other activities that are essential but are highly costly for food companies to have on an individual basis.

Cyber-attacks are a grim reality in today's business environment. Luckily, IT safety is given lot of importance in the UK food sector, which is well-resourced to tackle these attacks. Embracing new cloud-based management systems is an economical way to keep bad agents at bay.

Cyber-attacks can exist in different forms in the food sector, such as causing delivery disruption, recipe alterations, threat of tampering and accessing confidential information. Delivery disruption can disturb the food logistics system that moves products from one place to another, possibly stopping essential food reaching vulnerable consumers. Recipe alterations could occur by hacking into devices and sensors connected to a central system. Devices such as programmable logic controllers (PLCs) can be easily accessed remotely and then manipulated to create unwanted or disastrous recipes. Researchers have successfully demonstrated hacking PLCs in a water plant using a ransomware (Moore-Colyer, 2017). They managed to play with the monitoring systems by changing the chlorine levels in water. This showed that PLCs in food manufacturing operations can be easily manipulated by hackers; for example, an infant formula's nutrient content can be easily changed causing the infants consuming it to become ill.

Getting into confidential data is another area of interests for hackers. The recent Facebook scandal is a great example. In this scandal, a company named Cambridge Analytica harvested information from over 50 million user profiles through a quiz application, collecting data from not only the users taking the quiz, but also from the network of friends of these users (Granville, 2018). A joint study by McAfee and Science Applications International Corporation has shown that hackers are also penetrating the system to collect trade secrets, marketing strategies and other intellectual property to their own benefit (Shecterle, 2011).

Hackers issue threats of tampering with data crucial for the company. This practice can be adopted by cyber-activists, who usually disagree with a company's product or the way the company manufactures the product. They may hack into the company's system to tarnish its name, disrupt manufacturing, or spitefully alter processes and recipes. And based on the company's response, decide to launch the damage. Hackers may also give out threats for a ransom with regards to lost profit resulting from disruption of supply chain.

4.8. Augmented Reality

Augmented reality (AR) and virtual reality (VR) have recently gained importance in areas ranging from wine bottles to Domino's pizza (Cameron, 2018). Both Apple and Google have launched their AR platforms ARKit and ARCore, respectively (Shavel, 2019), while Facebook invested in its Oculus headset and Amazon launched AR shopping features (Leswing, 2019). In future, AR and VR will revolutionize people's lives.

AR and VR technology have also gained ground in the food sector. However, AR and VR technology development still remains high but food businesses are starting to see its potential benefits and to consider it as a valuable investment. Within the food sector, AR/VR technology is being deployed for staff training by human resources, to improve customer experience and to present the food products.

AR/VR technology is being used by human resources for staff training purposes. The traditional process of staff training is costly because it is heavily reliant on producing training materials, the quality of the trainer/team, space, time and travel. For example, to deliver a food safety training course across several geographical locations in a group company, a trainer can use AR technology to deliver the course from their preferred location but still give trainees a personal feel. This would save time, money and the need for a place to deliver the course (Bader and Jagtap, 2020). AR/VR technology comes at high investment but with high quality and finish. It can provide a detailed visualization for employees on how to undertake their everyday job and tasks in a safe manner. AR can also be effective in visualization of a recipe dish and serving sizes.

AR/VR has the capability to trigger all senses and engage the consumer to give them a branded feel. VR has found a lot of attention in the food and beverage sector. For instance, Boursin Sensorium has provided a VR-based experience which combined moving chairs, aroma and Boursin cheese samples. Patron tequila used a VR-based video to display behind-the-scenes tequila manufacturing (Johnson, 2015). Likewise, Innis and Gunn beer used a VR-based video to complement the taste of its beer (Dorsey, 2017). Kabaq, a tech-food start-up company, is making use of AR applications to enable food manufacturers to present food products in 3D and visualize the dishes using advanced scanning technologies. Grubhub, Dunkin Donuts, Subway and Magnolia Bakery are some of the companies attracted to Kabaq's AR application (Askew, 2018).

Innovative applications of AR/VR will soon become a reality and will form part of the visual experience for food and drink brands to educate, encourage and entice consumers to buy their products. The ability of AR/VR to provide additional information with visual experience and interaction would build the relationship between consumers, products and product contents, com-position and nutritional information. AR/VR can support food businesses by creating desirability among consumers, influencing consumer behaviour with regards to stimulating hunger, improving consumers' decision-making or com-fort about a product, consumers spreading positive word of mouth about their products and consumers helping themselves with the selection of higher-value products.

4.9. Cloud Computing

There is an increase in the number of incidents of food recalls or food alerts glo-bally. In the first quarter of 2019, food and drink recalls rose by 7.1% to 1041. This is reflected in the amount of recalls of unsafe food products in the UK, where a 40% increase was recorded for the year 2018 (Hughes, 2019). Overall, recalls can be very expensive and can harm a company's image. However, this can be avoided by having a system in place to identify issues at an early stage in the food supply chain. Rising food safety issues have encouraged food supply chain actors to deploy cloud-based sensors to monitor information on temperature, location and other parameters to analyse movement of the product at each stage of the food supply chain. This enables the supply chain actors to determine the time and place of any discrepancy. The US Food and Drug Administration has deployed a cloud-based system named OpenFDA to swiftly search for any query or pull any information through analysing massive amounts of data instantly (Kass-Hout *et al.*, 2016).

Cloud computing is becoming more and more important for the food sector. It is instrumental in gathering and analysing data for the whole supply chain, starting from the farms where crops/animals grow, the factories where they are processed, the warehouses that store finished products, to the transport system that ships them to retailers or consumers. Due to cloud computing technology becoming cheaper, the food industry can take advantage of this situation and gain an upper hand over their competitors.

4.10. System Integration

Currently, the food sector uses a range of software and platforms to per-form various functions, including operations, finance and human resources. However, these software and platforms do not always talk to each other or may be incompatible with other systems, making it impossible for the whole food supply chain to be connected. For example, if the human resources system and the operations system in the factory are working on different platforms and not sharing information between them then it might be difficult for the human resources system to know which employee is working in which department, while operations may find it difficult to plan their lines based on number of employees available at that time.

By having a standardized uniform data network system, various systems, departments and actors within the food supply chain can be integrated and remain connected. It allows seamless information exchange and the potential to have an automated food supply chain to be a reality. For instance, Epicor's Tropos Enterprise Resource Planning (ERP) tool is linked to the different departments and functions within a factory and performs various tasks, such as tracing product movement, production planning, stock control, finance, engineering, purchasing and logistics (Software Connect, n.d.). Having all the functions connected within the supply chain makes it easier to view each piece of information by all the actors in real time, resulting in cost and time saving and better decision-making (Hasnan and Yusoff, 2018). Therefore, we can conclude that system integration is a vital requirement for optimizing functionalities within the food sector.

4.11. Simulation

Simulation enables food supply chain actors to use real-time data to model physical supply chain systems or production and operations systems. This allows the supply chain actors to test, evaluate and optimize various parameters in a virtual environment before any real-life production/or changeover and decide on the best course of action (Rüßmann et al., 2015; Rostkowska, 2014). Currently, in the market a variety of simulation software is available, such as ARENA Simulation Software, TrackSYS, Tecnomatix and FlexSim. Simulation can lead to better decision-making in the early or start-up stages with regards to cost planning, minimizing downtimes and identifying any other production-related issues in advance. This has been demonstrated through a case study to design a new brewery with all the supply chain actions being simulated to understand, analyse and evaluate various operational approaches and situations in advance (Gathmann, 2018). Therefore, simulations in future will be utilized more widely in plant operations to leverage real-time data to reflect the actual world in a virtual model, which can include people, product, plant machinery, thereby reducing machine set-up times and increasing quality. Simulations can be created in 2D and 3D for virtual commissioning and for simulating cycle times, energy consumption or ergonomic features of a production floor. Application of simulations in production can reduce downtimes, changeovers and production failures during the starting-up phase. Also, making better decisions is possible with the applications of simulations.

4.12. Challenges of Implementing Industry 4.0 in the Food Sector

Implementing Industry 4.0 in the food sector comes with its challenges, such as cybersecurity, change management, employment, investment, to name a few.

4.12.1. Cybersecurity

The interconnectivity and digitization of systems is an important character-istic of Industry 4.0 as all devices are connected to each other over the Internet. However, this can result in cybersecurity issues such as danger of data theft, intellectual property and other business secrets (Woehl, n.d.). Therefore, there is a need for a strong security system to protect against hackers and other intentional/or non-intentional data breaches. Strong pol-icies and processes should be standardized and are needed to avoid or limit any data breaches.

4.12.2. Change management

Adopting changes is the recipe to success for most of food businesses. Industry 4.0 is also a big change on how businesses carry out operations through inte-gration of the physical and digital world (Geissbauer et al., 2014). However, while adopting a change of that scale, staff need support and to be provided with the necessary tools and skills in order to shift to an Industry 4.0 paradigm. Therefore, having staff at the core of the change allows staff engagement at all levels and becomes a key device for shifting to an Industry 4.0 paradigm. Hence, Industry 4.0 promoters and collaborators should join the team as change leaders, helping to embrace new technologies.

4.12.3. Employment

Industry 4.0 will heavily impact the traditional work culture as there is a need for staff to gain new Industry 4.0 based skills in order to be successful in these changing environments (Sima et al., 2020). Also, adoption of Industry 4.0 means phasing out of monotonous work performed by staff and replacing it with autonomous robots and machines running non-stop. Upskilling of staff is a requirement to keep up with the demands of Industry 4.0.

4.12.4. Investment

Industry 4.0 will come at a cost but will involve inexpensive IoT sensors or equipment fitted with Industry 4.0 solutions. Financing some of the Industry 4.0 solutions will affect a company's financial budget in the short term, but will reap enormous benefits in the long term. However, the food sector is con-tinuously changing as are consumer demands, so the high degree of flexibility needed may restrict businesses from investing in Industry 4.0 solutions. Also, collaboration between food industry experts and IoT implementers for per-forming data analytics, while resulting in better decision-making, can be costly (Vaidya et al., 2018).

References

ABB (2014) *Robotics. Food and Beverages Issue*. Available at: https://library.e.abb.com/public/89eb64523da1818048257ce500426115/ABB_robotics_1.14_low-res.pdf (accessed 28 July 2021).

Askew, K. (2018) Augmented reality and the future of food: 'There are many opportunities'. *Food Navigator*. Available at: https://www.foodnavigator.com/Article/2018/10/09/Augmented-reality-and-the-future-of-food-There-are-many-opportunities (accessed 21 August 2020).

ASTM (2012) *Standard Terminology for Additive Manufacturing Technologies*. ASTM F2792–10. ASTM International. Available at: https://www.astm.org/Standards/F2792.htm (accessed 28 July 2021).

Aung, M.M. and Chang, Y.S. (2014) Temperature management for the quality assurance of a perishable food supply chain. *Food Control* 40, 198–207.

Bader, F. and Jagtap, S. (2020) Internet of things-linked wearable devices for managing food safety in the healthcare sector. In: Dey, N., Ashour, A.S., Fong, S.J. and Bhatt, C. (eds) *Wearable and Implantable Medical Devices*. Academic Press, London, pp. 229–253.

Biji, K.B.R.C.N., Mohan, C.O. and Gopal, T.S. (2015) Smart packaging systems for food applications: a review. *Journal of Food Science and Technology* 52, 6125–6135.

Brantmark, H. and Hemmingson, E. (2001) FlexPicker with PickMaster revolutionizes picking operations. *Industrial Robot: An International Journal* 28, 414–420.

Buckenhüskes, H.J. and Oppenhäuser, G. (2014) DLG-Trendmonitor: Roboter in der Lebensmittel- und Getränkeindustrie (DLG trend report: Robots in the food and beverage industry). *DLG Lebensmittel* 9, 16–17.

Cameron, N. (2018) Domino's debuts augmented reality pizza ordering. Available at: https://www.cmo.com.au/article/649324/domino-debuts-augmented-reality-pizza-ordering (accessed 16 August 2020).

Cuthbertson, A. (2017) Ransomware attacks rise 250 percent in 2017, hitting U.S. hardest. Available at: https://www.newsweek.com/ransomware-attacks-rise-250-2017-us-wannacry-614034 (accessed 3 September 2020).

Dorsey, J. (2017) How augmented and virtual reality will reshape the food industry. Available at: https://techcrunch.com/2017/12/25/how-augmented-and-virtual-reality-will-reshape-the-food-industry/ (accessed 4 September 2020).

Duong, L.N.K., Al-Fadhli, M., Jagtap, S., Bader, F., Martindale, W., Swainson, M. and Paoli, A. (2020) A review of robotics and autonomous systems in the food industry: from the supply chains perspective. *Trends in Food Science and Technology* 106, 355–364.

Flynt, J. (2020) *A Detailed History of 3D Printing*. Available at: https://3dinsider.com/3d-printing-history/ (accessed 1 September 2020).

Gathmann, M. (2018) *From Quantity to Quality*. Available at: https://new.siemens.com/global/en/company/stories/industry/from-quantity-to-quality.html (accessed 11 September 2020).

Gebbers, R. and Adamchuk, V.I. (2010) Precision agriculture and food security. *Science* 327(5967), 828–831.

Geissbauer, R., Schrauf, S., Koch, V. and Kuge, S. (2014) *Industry 4.0 – Opportunities and Challenges of the Industrial Internet*. Available at: https://www.pwc.nl/en/assets/documents/pwc-industrie-4-0.pdf (accessed 7 September 2020).

Ghadge, A., Weiß, M., Caldwell, N.D. and Wilding, R. (2019) Managing cyber risk in supply chains: a review and research agenda. *Supply Chain Management: An International Journal* 25, 223–240.

Godoi, F.C., Prakash, S. and Bhandari, B.R. (2016) 3d printing technologies applied for food design: status and prospects. *Journal of Food Engineering* 179, 44–54.

Google Cloud (2016) Blue Apron: offering a better recipe for modern analytics. Available at: https://cloud.google.com/customers/blue-apron (accessed 13 August 2020).

Goyanes, A., Robles Martinez, P., Buanz, A., Basit, A.W. and Gaisford, S. (2015) Effect of geometry on drug release from 3D printed tablets. *International Journal of Pharmaceutics* 494, 657–663.

Granville, K. (2018) Facebook and Cambridge Analytica: what you need to know as fallout widens. *The New York Times*. Available at: https://www.nytimes.com/2018/03/19/technology/facebook-cambridge-analytica-explained.html (accessed 15 August 2020).

Greenberg, A. (2018) The untold story of NotPetya, the most devastating cyberattack in history. *Wired*. Available at: https://www.wired.com/story/notpetya-cyberattack-ukraine-russia-code-crashed-the-world/ (accessed 1 September 2020).

Hariharan, A. (2018) How to use data science to understand what makes wine taste good. FreeCodeCamp. Available at: https://www.freecodecamp.org/news/using-data-science-to-understand-what-makes-wine-taste-good-669b496c67ee/ (accessed 15 August 2020).

Hasnan, N.Z.N. and Yusoff, Y.M. (2018) Short review: application areas of Industry 4.0 technologies in food processing sector. *2018 IEEE Student Conference on Research and Development (SCOReD), Selangor, Malaysia, 26–28 November.* IEEE, pp. 1–6.

Hughes, N. (2019) Rise of the recall. *Food Manufacture*. Available at: https://www.foodmanufacture.co.uk/Article/2019/07/16/Why-food-safety-recalls-are-on-the-rise (accessed 26 August 2020).

Industrial Technology (n.d.) UK food industry warms up to robotics. *Industrial Technology*. Available at: https://www.industrialtechnology.co.uk/news--uk-food-industry-warms-up-to-robotics.html (accessed 13 July 2020).

Industry Europe (2020) Kuka delivers 'world's first fast food robot … probably' to Bionicook. Industry Europe. Available at: https://industryeurope.com/sectors/automationandrobotics/kuka-delivers-worlds-first-fast-food-robot-probably-to-bionicook/ (accessed 13 July 2020).

Jagtap, S. (2019) Utilising the Internet of Things concepts to improve the resource efficiency of food manufacturing. PhD thesis, Loughborough University, Loughborough, UK.

Jagtap, S. and Duong, L.N.K. (2019) Improving the new product development using big data: a case study of a food company. *British Food Journal* 121, 2835–2848.

Jagtap, S. and Rahimifard, S. (2019) The digitisation of food manufacturing to reduce waste – case study of a ready meal factory. *Waste Management* 87, 387–397.

Jagtap, S., Bhatt, C., Thik, J. and Rahimifard, S. (2019) Monitoring potato waste in food manufacturing using image processing and Internet of Things approach. *Sustainability* 11, 3173.

Jagtap, S., Bader, F., Garcia-Garcia, G., Trollman, H., Fadiji, T. and Salonitis, K. (2021a) Food logistics 4.0: opportunities and challenges. *Logistics* 5, 2.

Jagtap, S., Duong, L., Trollman, H., Bader, F., Garcia-Garcia, G. *et al.* (2021b) IoT technologies in the food supply chain. In: Galanakis, C.M. (ed.) *Food Technology Disruptions*. Academic Press, London, pp. 175–211.

Johnson, L. (2015) A bee's-eye view of the making of Patron Tequila: drones are filming seriously jaw-dropping VR video for brands. *International Society for Presence Research*. Available at: https://ispr.info/2015/05/05/a-bees-eye-view-of-the-making-of-patron-tequila-drones-are-filming-seriously-jaw-dropping-vr-video-for-brands/ (accessed 3 September 2020).

Kass-Hout, T.A., Xu, Z., Mohebbi, M., Nelsen, H., Baker, A. *et al.* (2016) OpenFDA: an innovative platform providing access to a wealth of FDA's publicly available data. *Journal of the American Medical Informatics Association* 23, 596–600.

Kondo, N., Monta, M. and Noguchi, N. (2011) *Agricultural Robots: Mechanisms and Practice.* Trans Pacific Press, Tokyo.

Kuswandi, B., Wicaksono, Y., Jayus, Abdullah, A., Heng, L.Y. and Ahmad, M. (2011) Smart packaging: sensors for monitoring of food quality and safety. *Sensing and Instrumentation for Food Quality and Safety* 5, 137–146.

Leswing, K. (2019) *Why Facebook and Amazon have joined the race to bring computing to your face.* CNBC. Available at: https://www.cnbc.com/2019/09/29/why-facebook-and-amazon-are-making-computer-glasses.html (accessed 11 September 2020).

Lipton, J.I., Cutler, M, Nigl, F., Cohen, D. and Lipson, H. (2015) Additive manufacturing for the food industry. *Trends in Food Science and Technology* 43, 114–123.

Mahadev, R.J. (2015) Using the Internet of Things (IoT) to digitize your factory. Cisco. Available at: https://blogs.cisco.com/manufacturing/using-the-internet-of-things-iot-to-digitize-your-factory (accessed 2 September 2020).

Moore-Colyer, R. (2017) Ransomware used to seize control of simulated water plant. Silicon. co.uk. Available at: https://www.silicon.co.uk/security/ransomware-water-plant-205417 (accessed 1 September 2020).

Pal, A. and Kant, K. (2018) IoT-based sensing and communications infrastructure for the fresh food supply chain. *Computer* 51, 76–80.

Palazzolo, K. (2015) *From the Ground to the Grocer, Barilla Makes Use of Cisco's Internet of Everything to Give Consumers Insight into the Journey of Their Food.* Cisco Newsroom. Available at: https://newsroom.cisco.com/press-release-content?articleId=1718599 (accessed 15 August 2020).

Periard, D., Schaal, N., Schaal, M., Malone, E. and Lipson, H. (2007) Printing food. *Proceedings of the 18th Solid Freeform Fabrication Symposium, Austin, Texas, August 2007*, 564–574.

Piller, F.T., Lindgens, E. and Steiner, F. (2012) Mass customization at Adidas: three strategic capabilities to implement mass customization. *SSRN* 1994981. Available at: https://ssrn.com/abstract=1994981 (accessed 28 July 2021).

Pinna, C., Ramundo, L., Sisca, F.G. and Angioletti, C.M. (2016) Additive manufacturing applications within food industry: an actual overview and future opportunities. *21st Summer School Francesco Turco, 2016*, 18–24.

Raza, M.Q. and Khosravi, A. (2015) A review on artificial intelligence based load demand forecasting techniques for smart grid and buildings. *Renewable and Sustainable Energy Reviews* 50, 1352–1372.

Rojko, A. (2017) Industry 4.0 concept: background and overview. *International Journal of Interactive Mobile Technologies (iJIM)* 11, 77–90.

Rostkowska, M. (2014) Simulation of production lines in the education of engineers: how to choose the right software? *Management and Production Engineering Review* 5, 53–65.

Rüßmann, M., Lorenz, M., Gerbert, P., Waldner, M., Engel, P. and Harnisch, M. (2015) Industry 4.0: the future of productivity and growth in manufacturing industries. *Boston Consulting Group* 9, 54–89.

Saldaña, E., Siche, R., Luján, M. and Quevedo, R. (2013) Computer vision applied to the inspection and quality control of fruits and vegetables. *Brazilian Journal of Food Technology* 16, 254–272.

Samuely, A. (2017) Domino's voice-activated ordering assistant tops half-million orders. Retail Dive. Available at: https://www.retaildive.com/ex/mobilecommercedaily/dominos-mobile-virtual-ordering-assistant-dom-tops-half-million-orders (accessed 14 August 2020).

Saravacos, G. and Kostaropoulos, A.E. (2016) Equipment for novel food processes. In: *Handbook of Food Processing Equipment.* Springer, Cham, Switzerland, pp. 605–643.

Sharma, A. and Jain, D.K. (2020) Development of Industry 4.0. In: Nayyar, A. and Kumar, A. (eds) *A Roadmap to Industry 4.0: Smart Production, Sharp Business and Sustainable Development.* Springer, Cham, Switzerland, pp. 23–38.

Shavel, T. (2019) *ARCore vs. ARKit: Which Is Better for Building Augmented Reality Apps?* Iflexion. Available at: https://www.iflexion.com/blog/arcore-vs-arkit (accessed 15 August 2020).

Shecterle, R. (2011) McAfee and SAIC say intellectual capital is new currency of choice for cybercriminals. *Industry Week.* Available at: https://www.industryweek.com/finance/article/

22011008/mcafee-and-saic-say-intellectual-capital-is-new-currency-of-choice-for-cybercriminals (accessed 15 August 2020).

Sima, V., Gheorghe, I.G., Subić, J. and Nancu, D. (2020) Influences of the Industry 4.0 revolution on the human capital development and consumer behavior: a systematic review. *Sustainability* 12, 4035.

Software Connect (n.d.) Tropos – An ERP system designed by Solarsoft Business Systems for manufacturing companies. Software Connect. Available at: https://softwareconnect.com/food-manufacturing/solarsoft-tropos/ (accessed 1 September 2020).

Splunk (n.d.) *Case Study: FreshDirect.* Splunk. Available at: https://www.splunk.com/view/SP-CAAACDB (accessed 13 August 2020).

Sun, J., Peng, Z., Zhou, W., Fuh, J.Y.H., Hong, G.S. *et al.* (2015) A review on 3D printing for customized food fabrication. *Procedia Manufacturing* 1, 308–319.

Tao, D., Yang, P. and Feng, H. (2020) Utilization of text mining as a big data analysis tool for food science and nutrition. *Comprehensive Reviews in Food Science and Food Safety* 19, 875–894.

Thymianidis, M., Achillas, C., Tzetzis, D. and Iakovou, E. (2012) Modern additive manufacturing technologies: an up-to-date synthesis and impact on supply chain design. 2nd International Conference on Supply Chains. Available at: https://cm.ihu.gr/logistics/images/logisticsdocs/icsc2012/fullabstracts/session_2/2_3_ICSC_12_THYMIANIDIS.pdf (accessed 28 July 2021).

Tirunillai, S. and Tellis, G.J. (2014) Mining marketing meaning from online chatter: strategic brand analysis of big data using latent dirichlet allocation. *Journal of Marketing Research* 51, 463–479.

Trivedi, M., Jee, J., Silva, S., Blomgren, C., Pontinha, V.M. *et al.* (2018) Additive manufacturing of pharmaceuticals for precision medicine applications: a review of the promises and perils in implementation. *Additive Manufacturing* 23, 319–328.

Trollman, H., Jagtap, S., Garcia-Garcia, G., Harastani, R., Colwill, J. and Trollman, F. (2021) COVID-19 demand-induced scarcity effects on nutrition and environment: investigating mitigation strategies for eggs and wheat flour in the United Kingdom. *Sustainable Production and Consumption* 27, 1255–1272.

Vaidya, S., Ambad, P. and Bhosle, S. (2018) Industry 4.0 – a glimpse. *Procedia Manufacturing* 20, 233–238.

Woehl, R. (n.d.) Cyber security threats to the food industry: consider the cloud. Global Food Safety Resource. Available at: https://globalfoodsafetyresource.com/cyber-security-threats-food-industry-consider-cloud/ (accessed 5 September 2020).

Zhou, H., Ding, Q. and Otto, J. (2013) The reality and prospect of fresh agricultural product supply chains in China. *International Journal of Applied Management Science* 5, 251–264.

Zhou, K., Liu, T. and Zhou, L. (2015) Industry 4.0: towards future industrial opportunities and challenges. *12th International Conference on Fuzzy Systems and Knowledge Discovery (FSKD),* 2147–2152.

Zou, Z., Chen, Q., Uysal, I. and Zheng, L. (2014) Radio frequency identification enabled wireless sensing for intelligent food logistics. *Philosophical Transactions of the Royal Society A: Mathematical, Physical and Engineering Sciences* 372, 20140209.

5 Revealing the Value of Resource Efficiency in the Food Manufacturing Sector

Sandeep Jagtap

5.1. Why Is Resource Efficiency Important for Food Manufacturing?

Improving resource efficiency is among the topmost priorities within the food manufacturing sector as it is increasingly concerned about the indiscriminate use of natural resources, negative environmental impacts, rising material costs and supply security. Therefore, using natural resources more efficiently has become indispensable in order to lead healthier lives, at the same time boosting the economy and respecting the limits of the planet. Being resource-efficient underpins the entire circular economy strategy, and it is essential to the growth of the food manufacturing sector. This chapter aims to stimulate innovation in the food manufacturing sector by addressing future technology trends and opportunities for growth in the circular economy through technological development and increased resource efficiency.

5.2. The Value of Resource Efficiency

The food sector is very resource-intensive and utilizes enormous amounts of raw materials (mostly food), energy and water. For instance, agriculture is one of the largest consumers of fresh water as well as consuming approximately 25% of the global energy during food production and distribution activities (United Nations, n.d.). Furthermore, one third of all food produced for human consumption goes to waste. As indiscriminate waste of resources continues to make them more stretched, the decision-makers in the food sector are now increasingly focusing on food waste, energy and water to underpin several of the United Nations Sustainability Development Goals (Rahimifard et al., 2017).

DOI: 10.1079/9781789248593.0005

The significance of resource efficiency in the food sector is tremendous and on par with any other industrial sector. Research carried out by Henningsson *et al.* (2004) demonstrated the value of resource efficiency in terms of environmental benefits as well as its financial benefits. It showed that with a little staff training and by probing the existing food operations, resource efficiency can be improved in an inexpensive manner. Therefore, the greatest financial savings in the food and drink sectors come in the form of raw material, energy and water savings while disposal of waste, treating effluents and reducing carbon emissions have very little impact.

In order to make food production and consumption more sustainable, all the actors (i.e. primary producers, manufacturers, distributors, retailers and consumers) within the supply chain must actively follow and adopt strategies based on resource efficiency and must emphasize reducing raw material wastage as well as reducing water and energy consumption. The actors should also pursue alternative solutions for improving resource efficiency. In this context, the next sections will describe the current situation and solutions that can be undertaken in order to reduce food waste, and water and energy consumption.

5.3. Food Waste Reduction and Management Solutions

It is widely accepted that preventing food waste is the best approach to addressing the issues associated with managing such food waste. Therefore, every actor in the food supply chain should give food waste prevention the topmost priority. However, the food supply chain will continue to generate food waste due to many reasons including errors during food production, manufacturing and processing. Sending food waste to landfill sites is considered the most convenient way and is common practice in several countries; but there are many opportunities available to transform it into other products. This can be demonstrated by adopting the food waste hierarchy. The food waste hierarchy comprises five steps, in order of preference: prevention, reuse, recycling, recovery and disposal as shown in Fig. 5.1. Due to the usefulness of this food waste hierarchy tool, it has been adopted by several food organizations and charities related to food to showcase the issue of food waste and the preferred option to tackle it.

Following the food waste hierarchy tool, the most preferred option is to prevent the waste of raw materials/ingredients by redistributing them to feed people or using them as animal feed. If this is not an option, then the food waste should be sent either to an anaerobic digestor for generating energy or composting to produce fertilizer. However, if this not an option, then it should be sent to an incinerator to recover energy; and the last option is landfill disposal. However, it is important to note that some food waste management solutions are controversial. For instance, in some cases the energy recovery through incineration is less economical due to the energy required to evaporate the high water content of food waste, while its use as a compost/soil enhancer is questionable (Lin *et al.*, 2013; UNEP and ISWA, 2015).

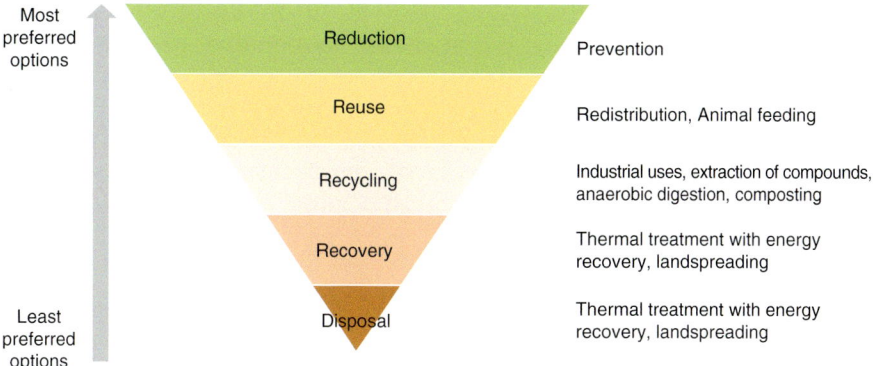

Fig. 5.1. Food waste hierarchy.

5.3.1. Prevention of food waste

Food waste happens at every stage of the supply chain and initiatives must be taken to avoid it from happening. In developed countries, the food waste is generated at the consumer level while in developing countries it happens during harvesting and distribution due to lack of cold storage or appropriate equipment. Prevention of food waste is applicable to edible foods; however, inedible parts, such as banana skins or egg shells are always considered as food waste.

FoodDrinkEurope conducted a survey of 29 companies and found that 80% of them are trying to identify the reasons for food waste generation. Addressing those reasons will eventually reduce the amount of food waste generated, thereby improving the production systems/processes (FoodDrinkEurope, 2014). Food waste can be prevented through adoption of waste management practices throughout the supply chain. Jagtap (2019) highlighted seven areas to prevent food waste generation as follows:

- quantifying/measuring food waste;
- engaging and raising awareness among all stakeholders within the supply chain about food waste;
- designing food products in such a way that they produce less waste;
- processing and packaging with a view to food waste minimization;
- selecting a new product range/lines to reduce food waste;
- accurate forecasting and demand; and
- reconfiguring processes to ensure products and information flows are seamless.

Supply chain actors such as retailers could wield considerable influence in order to reduce food waste. For instance, incorrect forecasts or orders from retailers could lead to overproduction of partly or completely finished products which may soon expire or be out of date. Also, as per the manufacturer-retailer contract agreement, if the food products on retailers' shelves that are not sold or are nearing the end of their life can be sent back to the manufacturers, this creates more waste (Stuart, 2009). Furthermore, a short window given by the retailers to the manufacturers to amend last-minute sales orders also creates food waste. However, this situation can be addressed by allotting extra time to complete the orders, achieved through collaborative communication.

In order to reduce food waste, some food manufacturers adopt standard processes, which are shared with all or most of their products (e.g. boiled rice), and then complete 'special operations' to give the products their specific individuality (e.g. adding sauces) (Darlington and Rahimifard, 2006). Other ways to prevent food waste are through earmarking the rejected food products with imperfect visual quality (e.g. wonky fruits and vegetables, fruits with discolouration and spots on the skin) for different usage. They could be peeled and used in convenience foods, for example. The food manufacturing processes and systems, including equipment, must be upgraded in order to reduce food waste; for example, reduction in excessive trimming of food products and minimization of food waste due to overprocessing or overcooking. Likewise, manufacturers could explore design of products or packaging in order to improve shelf-life of food products. However, shelf-life is also dependent on factors such as good manufacturing practices (GMP), effective hazard analysis and critical control points (HACCP), raw material quality, processing stages, packaging material, distribution conditions, temperatures at storage, product design, final use and the consumer.

As already discussed, every stage of the food supply chain will generate some kind of food waste, and when preventing the waste may not be an option, reusing it is the next best option. Reusing surplus food can be achieved through redistributing it for human consumption or feeding to animals.

5.3.1.1. Redistribution

The overproduced or surplus food produced by the supply chain actors may be redistributed for human consumption through food charities or food banks; although many food organizations and stakeholders do not directly support food banks or charities due to the legalities involved in their supplier-manufacturer-retailer agreement. However, a number of charitable organizations participate in food redistribution at the national level, such as, in the UK, FareShare, FoodCycle, Olio and Plan Zheroes. Efforts are undertaken to redistribute surplus food from retailers to local food organizations through an IT-based system (Buksti et al., 2015), including products that are fit for human consumption but are rejected due to issues such as wrong labelling, wrong weights, nearing expiry date, wrong specification and mechanical damage (Bilska et al., 2016).

Redistribution of food for human consumption must be prioritized as it can provide both social and environmental benefits to the food donor (Giuseppe et al., 2014). But, at the same time, in some cases, redistributing food to people in need cannot be economically feasible when compared with the options available further down the food waste hierarchy (Reynolds et al., 2015).

5.3.1.2. Animal feed

If redistribution of food for human consumption is not an option, it can be diverted to produce animal feed. Catering or finished food products such as ready meals, sauces etc. may have high salt content or be contaminated with animal by-products, which is banned in some countries (DEFRA, n.d.; Gov.uk, 2014). Some food products, such as breakfast cereals and breads, are rich in carbohydrates and can be used as animal feed (Parfitt et al., 2016). Fruit and vegetable waste is processed using various drying techniques such as microwave ovens, and pulse combustion drying, to produce nutritionally rich and safe animal feed (San Martin et al., 2016).

Over the past few years, due to intensive animal farming, there has been much use of grains such as corn and soybean for animal feed rather than grass and farm or food waste. Feeding pigs on food waste saves at least 20 times more carbon emissions (Stuart, 2009) and lowers the negative environmental impact as it is not sent to an anaerobic digestor for composting (Salemdeeb et al., 2017).

5.3.2. Recycling

Surplus food which cannot be redistributed for human consumption or animal feed can be processed to extract compounds, or it can be sent to an anaerobic digestor to recover energy and may be composted.

5.3.2.1. Compound extraction

Several valuable compounds can be extracted from food waste and can be used in various applications (Ravindran and Jaiswal, 2016). But, some of the extraction processes are expensive and involve complex processes and technologies, and some of the technologies are outdated and may require scaling up, which may add further challenges (Galanakis, 2015; Matharu et al., 2016). Another challenge is industry know-how and shortage of skilled staff, and with the variable and seasonal nature of food waste, this may create challenges for the extraction of valuable compounds (Lin et al., 2014). To address the variation and seasonal nature of food waste, it is more suitable to extract compounds from food waste during the early stages of the supply chain since it is homogeneous and there is a regular supply (Girotto et al., 2015). However, every extraction process will generate some kind of waste, which may need further treatment.

A review of technologies and processes to extract valuable compounds was presented by Waldron (2007). Animal fat can be extracted from animal by-products to produce animal feed, soap, fuel and other products (Meeker, 2006). Essential oils, colourings, flavourings and aromas can be extracted from fruit and vegetable waste (Morawicki, 2012), whereas materials extracted from citrus waste and wheat bran can be used in the production of new foods (Fava et al., 2013). A wide range of products can be extracted from the food waste such as emulsifiers (Gould *et al.*, 2016), succinic, lactic and fatty-acid based plasticizer (Pleissner *et al.*, 2016).

5.3.2.2. Anaerobic digestion

Anaerobic digestion is a process in which biodegradable matter such as food waste is broken down by microorganisms in the absence of oxygen to generate biogas and a residue termed as digestate. Anaerobic digestion is a favourable technology compared to other traditional disposal methods such as landfill, incineration and composting for managing food waste. Considering the damaging environmental effects of landfilling, incineration and composting of food waste, anaerobic digestion is considered to be economical for renewable energy production and waste treatment of this high-moisture and energy-rich material (Romero-Güiza *et al.*, 2016). During anaerobic digestion, microorganisms process organic material and convert it into biogas, carbon dioxide, other gases (hydrogen and hydrogen sulphide) as well as by-products such as nutrient-rich manure, which is used as fertilizer (Xu *et al.*, 2018). Most of the anaerobic digestion plants in western countries utilize agricultural and food waste for electricity generation. Food waste is considered perfect raw material due to its high moisture content, leading to easy biodegradability for biogas generation (Sen *et al.*, 2016; Chiu and Lo, 2016). Anaerobic digestion is considered to have a more positive environmental impact than composting (Fisher *et al.*, 2013).

5.3.2.3. Composting

In composting, bacteria in the presence of oxygen, transform organic waste into a nutrient-rich soil conditioner usually termed as compost. Composting at household or industrial level is usually carried out in vessels such as containers, silos, bays, tunnels and halls (WRAP, 2016). Composting is considered an economically beneficial waste management solution as it is an inexpensive process which eliminates the need to pay disposal fees. It reduces the quantity of waste by at least 40% as well as destroys most of the pathogens due to high temperatures during the process (Schaub and Leonard, 1996).

5.3.3. Recovery

When recycling is not an option, food waste could be processed in such a manner that it generates maximum benefit by recovering value from it. The most common method to recover value is through thermal treatments with energy recovery and land spreading.

5.3.3.1. Thermal treatment with energy recovery

Thermal treatments with energy recovery include processes such as incineration, gasification and pyrolysis; however, they differ in terms of the temperatures and material produced (Arvanitoyannis, 2010). Energy in the form of heat or electricity is recovered through these waste treatments (Kwak *et al.*, 2006). Moreover, the residues such as bottom ash, may be beneficial in other industrial applications (Ahmed and Gupta, 2010; Brunner and Rechberger, 2015). In the incineration process, the waste is burnt between 870 °C and 1200 °C, which converts solids and liquids into gases. It is further broken down into simpler molecules which react with oxygen to produce CO_2, H_2O, CO, nitrogen oxides and other compounds; the leftover is ash and slag (Arvanitoyannis, 2010). Pyrolysis is an incineration process generally carried out under pressure and at temperatures more than 430 °C in the absence of oxygen. The main gases produced are CO, H, CH_4 and other hydrocarbons with residues such as liquid and coke. In the gasification process, the temperature is higher than 700 °C, the waste in the presence of controlled O_2 and steam produces syngas, which is a mixture of CO and H and residues ash and char (Arvanitoyannis, 2010).

Although not all the processes described above are as efficient as coal-fired power stations, and produce ashes and pollutants that are harmful to humans and have a negative impact on water, soil and air (FAO, 2013), those processes substitute for the burning of fossil fuels, and can be considered a renewable source. They considerably reduce the amount of waste and eliminate potential pathogenic microorganisms and viruses (Brunner and Rechberger, 2015). Apart from being an expensive method (Thi *et al.*, 2015), it also has a negative environmental impact due to gas emissions and negative social impact due to odour, noise, dust and traffic (DEFRA, 2013).

Although the energy recovery from thermal treatments of food waste is not suitable due to the high water content, the results yield savings by compensating for the use of fossil fuels and reducing disposal costs (Caton *et al.*, 2010). Thermal treatments are the second most common method to address global waste after landfill and open dumping (Arvanitoyannis *et al.*, 2006).

5.3.3.2. Landspreading

Landspreading involves the practice of spreading organic waste such as food waste onto the land to increase the nutrient level of the soil. This practice enhances the physical, chemical and biological qualities of the soil, reducing dependency on manufactured fertilizers (Environment Agency, 2013). At the farm level, farmers can spread food waste, usually inedible parts of plant-based material, onto their land without needing to store or transport it. If they decide to spread animal-based material, before doing so they must check the proximity to groundwater sources as it can be contaminated with harmful microorganisms (Environment Protection Agency, 2004)

5.3.4. Disposal

Disposal of food waste is the final option as per the food waste hierarchy, which should always be prevented. It involves both thermal treatment without energy recovery and landfilling.

5.3.4.1. Thermal treatments without energy recovery

Thermal treatments without energy recovery involve burning the waste in the open air, i.e. without energy recovery. The by-products of this treatment result in char, which is used in industrial and domestic applications as a fuel and soil enhancement (Lohri *et al.*, 2015). Although this process reduces the volume of waste and hazardous materials are destroyed (Brunner and Rechberger, 2015), the process results in loss of heat and the gases emitted in the process are dangerous and contribute to the greenhouse effect. Hence, due to no benefits and the risk of air pollution, this process is highly discouraged (Hoornweg and Bhada-Tata, 2012). This practice of thermal treatment of waste is more visible in developing regions than developed regions (Guendehou *et al.*, 2006) and food waste is generally mixed with other waste and burnt in the open air without energy recovery (Caton *et al.*, 2010).

5.3.4.2. Landfill

Landfill is a waste disposal process wherein the collected waste is deposited onto, or into, a parcel of land (Environment Agency, 2010). Landfill has substantial negative environmental impacts and minimal benefits, irrespective of whether the waste is buried or not. Decomposition of organic matter releases methane and carbon dioxide, which heavily contribute to the greenhouse effect. Nevertheless, in some cases these gases can be captured to produce energy in well-managed landfill sites (Emkes *et al.*, 2015). However, other toxic components produced during this process can contaminate air, water and land, and are hazardous to human health (FAO, 2013).

Fortunately, landfilling is the least preferred option in the waste hierarchy, and should be prevented due to its negative impact on the environment. Landfilling waste in a controlled environment in a well-managed landfill site is a preferred option to openly dumping or fly-tipping. In addition, directing food waste into the sewer is as bad as landfilling (WRAP, 2017), and therefore is the last option on the waste hierarchy.

5.4. Energy Optimization and Minimization Practices

In the last 150 years, the global demand for energy has increased tremendously to fulfil the needs of industry and the growing population. Currently, we are fully reliant on ample and uninterrupted supply of energy for both industry and domestic applications. It is a key resource across all sectors of modern economies. However, it is estimated that by 2030, the demand for energy will rise by 50% (IEA, 2018), mainly due to developing countries like China and India, as shown in Fig. 5.2.

When it comes to energy consumption, the high-temperature processing industries such as metal, bricks, cement, glass and pottery dominate, while low-temperature processing is dominated by food, drink and tobacco industry (Pathare *et al.*, 2019). Energy consumed for space heating and lighting

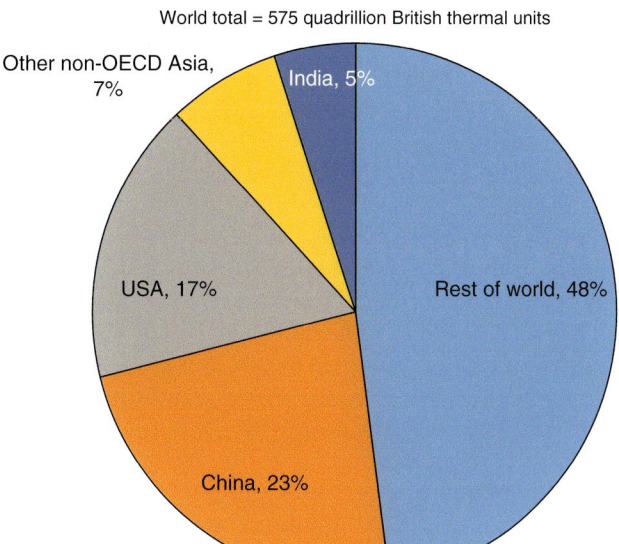

World primary energy consumption 2015

World total = 575 quadrillion British thermal units

Fig. 5.2. World primary energy consumption 2015 (EIA, 2017).

are dominated by the engineering sector; compressed air is dominated by the publications sector (Griffin *et al.*, 2016); and refrigeration with chemicals is dominated by the food and drink sector. Minimization of energy consumption in the past has been encouraged in manufacturing as a means to save money. The cost-effective, energy-efficient benefits in the range of 10–20% are possible through commercially existing technologies for energy-intensive industries such as paper, chemicals, glass, steel and cement manufacturing (Vallack *et al.*, 2011). In some cases, approximately 20% reduction in energy costs is equivalent to 5% growth in sales (Carbon Trust, 2018). Therefore, previous energy-minimization attempts were driven by profitability (Ptasinski *et al.*, 2006) but future attempts will be driven by relaxation in taxation. A study conducted by Worrell *et al.* (2009) shows that the manufacturing sector can gain from energy efficiency and productivity.

In this context, the food sector is also a major consumer of energy. Globally, the food sector consumes around 200 EJ per annum, of which 45% is utilized in processing and distribution activities (FAO, 2017). The food sector is heavily reliant on fossil fuels and generates considerable greenhouse gas emissions. Increasing population size will lead to higher food demand and it is estimated to increase by 60% by 2050. Hence, energy efficiency in the food sector is an important issue due to rising demands and costs of electricity and fossil fuels as well as cut-throat competition, which is reducing profit margins. Therefore, to address this business challenge, as well as negative environmental impacts associated with the food sector, and to make it more sustainable, energy usage needs to be optimized and minimized during food manufacturing. However, some organizations have positively embraced environmental sustainability with regard to their products and operations and consider it a business option rather than a challenge (Grayson and Hodges, 2017).

The amount of energy required for industrial processing of food is associated with the economy of the country (Monforti-Ferrario *et al.*, 2015). The authors further state that in low-income countries, industrially processed food accounts for almost 30% of food, while in high-income countries it accounts for approximately 98%. However, processing of food requires energy and the more processed the food the higher the energy required (Monforti-Ferrario *et al.*, 2015). The energy consumed per unit by various processed food products varies, i.e. 1 kg of processed beef will have a different energy requirement to 1 kg of processed chicken. In some cases, even the energy consumed by similar food products may differ depending on the country where they are processed. With such a variation in energy needs, it is challenging to specify tangible energy improvements at the sector level. Even the production structure can be an important aspect, i.e. a reduction in the industrial energy consumption per food product per unit could be the result of improvement in efficiency, or a move towards production of less energy-intensive food products.

The complex nature of the food makes energy efficiency a tougher task. Nevertheless, the energy consumption in food manufacturing in recent years is decreasing gradually, both in regard to total consumption as well as in regard to energy consumption per unit of production value (Jagtap and Rahimifard, 2017, 2019). This demonstrates that the food industry is pursuing the energy efficiency pathway, i.e. producing more with less energy. As per EIA (2017), the food industry is considered a less energy-intensive sector and the direct energy cost accounted for 3% of the total production cost for an average company. Therefore, company managements search for cost optimization opportunities in other areas before considering possible savings through energy usage minimization.

5.4.1. Technological and process optimization

Energy consumption in the food industry can be minimized through process optimization; however, specific energy costs related to chilling, freezing, cooking etc. cannot be prevented, but the way in which the process is executed can produce significant savings (Jagtap *et al.*, 2019). Without radical changes in the policy or behavioural attitudes, only 5–7% of energy reduction is estimated to be attainable by 2030 (Altmann *et al.*, 2010).

Analysing processes within food manufacturing reveals that cooling and freezing is responsible for 30% of electricity consumption, which is comparatively higher than other manufacturing sectors (Altmann *et al.*, 2010). As per Eichhammer *et al.* (2009), cooling and freezing, by 2030, will account for 30% of long-term technical possibilities for energy savings and 20–26% of cost-effective savings possibilities. Table 5.1 demonstrates how various researchers attempted or achieved energy-saving opportunities within food manufacturing. One of the analyses conducted by Thollander and Palm (2013) highlighted that in the low energy-intensive small and medium enterprises, up

Table 5.1. Energy-saving opportunities in food manufacturing.

Abdelaziz *et al.* (2011)	Energy-saving opportunities from using high-efficiency motors, use of combined heat and power, intelligent and efficient equipment and lighting, voltage optimization, which can be integrated into energy management systems
Thollander and Palm (2013)	Focuses on individual production processes, which are important to the food sector such as decomposition, mixing, cutting, joining, coating, forming, heating, melting, drying/concentration, cooling/freezing and packing. Focuses on support processes such as lighting, compressed air, ventilation, pumping, space heating and cooling, hot water and internal transport
Kaminiski and Leduc (2010)	Identified most essential systems and processes where substantial energy-efficiency improvements can be attained: steam systems, motor and pump systems, compressed air system, process cooling and refrigeration, heating and lighting of premises. Also, established a wide range of possible practical improvements for the named systems such as leak maintenance, proper motor sizing, and condensate return systems
Burfoot *et al.* (2004)	Assessed the energy savings by substituting ambient cooling with localized air delivery in chilled food production
Damour *et al.* (2012)	Showed the benefits of optimizing defrosting systems
Miah *et al.* (2014)	Opportunities provided by heat recovery optimization in complex production lines for a multi-product confectionery factory
Krasulya *et al.* (2014)	Reported tangible energy savings obtained from the introduction of ultrasound-based tools in processes such as emulsification, filtration, tenderization and functionality modification

to 70% of energy is utilized in support processes while in larger energy-intensive factories, up to 85% of energy is spent on production processes.

5.4.1.1. *Plant system improvement*

Although energy savings are recognized at process level, the real savings achieved at plant level are not straightforward to analyse. The savings depend on various factors such as share of each individual processes in the factory budget, the factory design and its age and maintenance status (Jagtap and Rahimifard, 2018). Furthermore, different improvements to the same process do not always add value, and improvement to one process is likely to influence the energy consumption of other processes.

Muller *et al.* (2007) developed a linear regression model to aid factory management within the food processing industry with the practical implementation of effective energy-efficient practices by tracking energy-saving opportunities. The method needs information on design data and operational data for a certain period to realize the most suitable effective changes.

The model was tried in a Swiss food factory and the outcome suggested improvements to air compression and vacuum production (Muller *et al.*, 2007). Kannan and Boie (2003) suggested the idea of combining energy audit and energy balance sheets at a small-scale German bakery, which resulted in permanent energy savings of 7% at little or no cost. Analysing processes are challenging to extend beyond the individual factory since the outcomes are dependent on specific factors to do with the factory itself (Kannan and Boie, 2003). However, Seck et al. (2013) established a full bottom-up approach for energy savings resulting from heat recovery and heat pumps throughout the whole of the food and drink sector in France.

5.4.1.2. In-plant energy usage

Reusing the by-products produced during food production, such as in anaerobic digestion, could be utilized for energy generation leading to energy savings (Hall and Howe, 2012). Biogas obtained from the anaerobic digester plant through decomposing biodegradable by-products could be used for generation of heat and power or as fuel for ovens and boilers. Jensen and Govindan (2014) demonstrated the feasibility of using biogas to generate power and heat from a Danish bakery producing 20,000 tonnes of by-products per year. Their study showed that return on investment is mostly positive. However, regional or national policies may influence their approach (Jensen and Govindan, 2014).

5.4.1.3. Renewable opportunities in food processing

The food processing industry is a moderately energy-intensive sector and is comprised generally of small and medium-size enterprises. In those cases, the industry may not be looking for practical energy cost savings but at the option of greening their energy source, in exchange for additional value through governmental incentives.

Mekhilef et al. (2011) studied the use of solar energy in industry and its possible applications when a constant flow rate of moderate heat between 80 °C and 120°C is required. It can be successfully used for washing, cleaning, cooking, pasteurizing, sterilizing, distillation, hydrolysing, evaporation, polymerization and extraction activities in food processing. Solar energy, if continuously available, could be a relevant and inexpensive substitute for non-renewable energy sources (Mekhilef et al., 2011). Twelve per cent of energy consumption in the food sector is utilized for drying processes; therefore Pirasteh et al. (2014) discussed the idea of increasing solar-based drying irrespective of geographical settings and existing solar technology. Muller et al. (2007) discovered that within the brewing industry in Germany almost 30% of the thermal energy required for drying could be obtained through solar energy. Even though use of solar energy is less prevalent, it could be utilized for cooling purposes in the food industry, and Best et al. (2013) have demonstrated how this resulted in 19% of electricity savings in a Mexico-based meat-processing factory.

Attwood (2008) reported that McCain Foods in the UK will be utilizing wind power to generate energy, which will reduce the site energy bills by 60% and also lead to a reduction of approximately 20,300 tonnes of carbon dioxide greenhouse emissions. The International Renewable Energy Agency estimated that food sector SMEs could have almost 50% of their energy needs supplied through renewable technologies, including biomass, wind power, solar thermal systems, and geothermal and heat pumps (IRENA, 2015).

5.4.2. Energy efficiency in food transportation

Given the long distances food products travel these days along with the general need for chiller/freezer systems, food transport is a vital component in the food–energy balance. Enhancing the energy performance of food transportation systems and reducing or improving the amount of transportation itself are two ways to improve energy efficiency. However, food transport not only needs energy to move vehicles, but also one third of food transported using roadways needs refrigeration systems (Tassou *et al.*, 2009).

5.4.2.1. Enhancing refrigeration

Significant amounts of fruit and vegetables travel long distances before reaching consumer tables, either by road, sea or, in some cases, air. Tassou *et al.* (2009, 2010) have analysed road transport and reviewed the key existing technologies for the in-vehicle refrigeration systems and their potential improvements. They established that commonly found vapour-compression refrigeration cycle-based systems are oversized to provide a significant operational margin. The auxiliary diesel engines, which are nearly always fitted to provide back-up in case of refrigeration system failures, are rarely used in reality. In addition, the large amount of thermal energy emitted from the diesel exhaust could also be recovered and utilized to power the thermally driven refrigeration system as well as power generation through thermo-electrics or turbo generators. Furthermore, Tassou *et al.* (2009) claim that heat generated as a by-product from the engine of vehicles could be enough to power refrigeration systems in usual out-of-town driving conditions, while extra energy would be needed only for in-town driving (Tassou *et al.*, 2010). The 12% of energy consumed in the transportation phase is associated with retail. According to Tassou *et al.* (2011), up to 60% of energy consumption is related to chilling and refrigeration in small shops with a size below 1400 m², while in larger shops or supermarkets, lighting, heating and other equipment consume most of the energy. Therefore, choosing the right refrigeration systems and appropriate design of food display cabinets are the most favourable and feasible energy-saving opportunities in food shops (Tassou *et al.*, 2011).

5.4.2.2. Reduction in transportation requirements

Reduction in the total distance travelled by food is another way of decreasing the amount of energy consumed by transportation. It can be achieved in two ways: firstly, by optimizing transport supply; and secondly by reducing transport demand. Intelligent transport systems can help with optimizing food transport through telematics, communication and control and automation technologies (Gattuso and Pellicanò, 2014). From the transportation demand viewpoint, reducing the overall energy impact of food products can be realized by using locally grown food. It is generally considered that food travelling longer distances embeds more energy than locally grown food; however, many studies have shown that the issue needs to be sensibly analysed on a case-by-case basis. For example, a study conducted by Blanke and Burdick (2005) demonstrated that locally grown apples are less energy-intensive when compared to the same quantity imported from New Zealand. Schlich and Fleissner (2005) adopted a Life Cycle Analysis approach to compare the energy consumption of fruit juice and lamb from Europe against similar products imported from South America and Australia and found that, although food miles were an issue, there is a stronger association between energy content and the size of the business.

5.4.3. Energy in food packaging

Materials such as metal, glass, plastic, ceramics, paperboard and paper are regularly used for the manufacturing of food packaging. The safety of these materials is dependent on ensuring that during contact with food there is no passage of unsafe chemicals and that the materials are safe for consumers. Almost 10.7% of the embedded energy in European Union food consumption is associated with packaging. Improved packaging sustainability must not compromise food quality (Russell, 2014), while good and long-lasting packaging is important for reducing food waste and extension of food shelf-life, thus lowering the overall energy consumption.

5.4.3.1. Developing optimal packaging

Barlow and Morgan (2013) have studied the trade-offs present in the food packaging development focusing on pros and cons of single-layer packaging against multi-layered packaging. The former is better for recycling and the latter usually ends up in landfill. They further suggested that optimization and energy reduction can also be addressed by looking into secondary (boxes, trays) and tertiary packaging (pallets, shrinkwrap).

5.4.3.2. New packaging materials

Bioplastics obtained from biological materials are still being researched and pose a threat by their wide usage (Peelman et al., 2013), such as brittleness, low-melt strength, thermal variability, problematic heat insulation and high water vapour and oxygen penetrability. This restricts their usage as packaging for short-shelf-life products (e.g. fruits and vegetables) and in some cases even the long-shelf-life products (e.g. pasta and chips), since they do not require very high oxygen or water-barrier properties. Puligundla et al. (2012) discussed intelligent food packaging, which allows tracking and converse information about the state of packed food, thus ensuring optimal food consumption and reducing food waste and associated energy waste.

5.5. Water Optimization and Minimization Practices

Water is an important constituent in food and drink manufacturing processes and widely consumed for various activities such as washing, processing, blanching, steaming, boiling and cleaning. For instance, the primary production of food requires huge amounts of water and more than two thirds of all freshwater abstraction globally (up to 90% in some countries) goes towards food production (Kirby *et al.*, 2003). Moreover, the sector is also facing the increasing cost of mains water and is aware of the fact that good water-management practices can substantially reduce water costs. One of the key features of the food and drink sector is demand for high-quality potable water in the majority of its processes to meet stricter regulatory standards and ensure the manufacturing of safe food products. For this reason, the food industry is dependent on mains water for most of its requirements. However, in the last decade there has been a drop in the usage of mains water in the food sector and an increase in water abstraction from underground sources and rivers. This sudden shift is due to the increase in water prices from the water companies (IGD, 2007).

As per Ellis *et al.* (2001), process water (31%) is associated with the largest water consumption in the food sector. The study also demonstrated that most water-demanding processes in food processing varied from product to product and specific operations. For instance, in sugar production it has been revealed that approximately half of the water consumed is for cooling, while 20% or less water is consumed for processing (Ellis *et al.*, 2001). In contrast, meat processing and fruit preservation consumes about 60% of water as process water (David, 1990).

5.5.1. Techniques and technologies for water reduction

Globally, the food sector is one of the largest consumers of water. Water and wastewater treatment costs are increasing every year. The cost of water as a resource and discharge of water effluent form a substantial part of the food manufacturers' operating costs. Moreover, stricter regulations on emission of water mean that the food sector can benefit from minimizing its water usage and volume of effluent discharge (WRAP, 2013). With this perspective, a review of the best available techniques and technologies to minimize and eliminate water usage in food manufacturing is discussed in the following sections.

Table 5.2 depicts the main functions that use water in the food-and-drink sector. The journey of water begins from the time the raw materials enter the factory premises until it is processed into finished products. The water journey provides an idea of how important the water is and its role across all the processes in food manufacturing, and hints can be gleaned as to where potential improvements in efficiency and savings are likely. As illustrated in the table, water is utilized for numerous operations at all stages of food processing, including washing and cooking food, food preparation and cleaning.

Table 5.2. Key water-use processes in food manufacturing (Mistry *et al.*, 2007).

Product preparation	Processing	Preservation	Packing	Equipment cleaning	Site facilities
Washing	Mixing	Freezing	Canning	Cleaning in place	Site maintenance
Cleaning	Steaming	Heating	Bottling	Rinsing	Vehicle washing
Peeling	Pumping/ transferring	Boiling		Washing	Staff facilities
Cutting	Water in product			Sterilization	

There are various techniques and technologies available for water efficiency and those can be categorized into four areas as shown in Table 5.3:

- efficiencies – no-, or low-, cost solutions;
- efficiencies – technical solutions;
- reuse and recycling – technical solutions;
- effluent reduction and treatment – technical solutions.

5.5.1.1. Efficiencies – no, or low-cost solutions

TRAINING. Training includes staff members training on techniques for sensible water usage and encouraging them to be more water-efficient (Henningsson *et al.*, 2004). Training could be achieved through regular meetings to update staff on best practices and delivering training sessions to demonstrate to staff how to operate and clean equipment correctly (WRAP, 2013).

It further states that the aim of effective training is to ensure that staff members are aware of the subject of resource efficiency and are utilizing best practices to minimize water wastage. This approach can only be effective if it is part of a continuous improvement plan on resource efficiency rather than a one-off session.

GOOD PRACTICE. Good practice involves creating procedures of performing activities that confirm that the current practices to reduce water are being effectively used (WRAP, 2013). Good practice includes using equipment efficiently, such as turning the hosepipes or taps off when not in use and ensuring that equipment is correctly bunded. Good practice methods could be applied to each stage of the business and linked with training sessions to ensure that employees are operating equipment efficiently.

METERING. According to WRAP (2014), using water meters is the first step towards minimizing consumption of water in a business. Sub-metering allows tracking and recording water usage in different departments within a factory and can be linked to data-loggers to relate its usage with production and shift-cleaning water usage (WRAP, 2014; Jagtap *et al.*, 2018). It is also useful to determine baseline data and to detect any leakages during breakdowns or site closures.

Table 5.3.   Key water-saving techniques and technologies in food manufacturing (Mistry *et al.*, 2007).

Short-term (no-/low-cost)	Medium- to long-term (medium to high cost)		
Efficiency management	Efficiency – technical solutions	Reuse and recycling – technical solutions	Effluent reduction/ treatment – technical solutions
Training Good practice Metering and online analysers	'Water pinch' (process integration) CIP optimization systems Mechanical-seal water management Floor washers Closed-transfer equipment	Rainwater harvesting Counter-current rinsing Mechanical-seal water management Membrane filtration Ozone/UV Vehicle washers	Pigging Slurry de-watering/ drying Electro-coagulation Anaerobic digestion Online analysers Dosing equipment Closed-transfer equipment UV/ozone Sand filters, dissolved air flotation

5.5.1.2. Efficiencies – technical solutions

WATER PINCH.   Water pinch is a strategic tool which allows analysing water networks and reducing water usage in processes. It helps to realize and optimize water reuse, water regeneration and effluent water-treatment opportunities through a set of advanced algorithms (Lu *et al.*, 2018). Furthermore, it helps minimize the losses of both feedstock and valuable components that are present in effluents. Between 20% and 60% of reductions in effluent flow can be achieved through water-pinch technology (Mistry *et al.*, 2007). For example, in a corn refinery using the pinch analysis, the daily water consumption decreased from 1800 m³ to 1235 m³ if chemical oxygen demand (COD) was taken into account, and to 1152 m³ if total dissolved solids (TDS) were considered (Bavar *et al.*, 2018). Paudel *et al.* (2017) demonstrated that 25% of water savings could be achieved through water-pinch analysis in a mushroom-canning process. Other than reducing water-volume charges, this technique, when used in combination with dosing equipment and online analysers, helps in reducing biological oxygen demand (BOD) and chemical oxygen demand (COD) as charges are also dependent on pollution load (Mistry *et al.*, 2007).

CLEANING IN PLACE (CIP) SYSTEMS.   CIP systems provide extremely effective means of cleaning huge tanks and vessels that need regular cleaning. CIP systems use high-pressure cleaning in combination with spray balls, nozzles or rotating heads (Amazonas, 2009). As CIP is an automated system (i.e. no humans involved), it can use strong detergents and chemicals for effective cleaning with optimal use of water. Palabiyik *et al.* (2015) proposed a two-step CIP protocol that reduced the energy consumption by almost 40% without affecting the cleaning efficiency. If used in conjunction with partial solvent recirculation, these systems can improve washing efficiency by 90% (Mistry *et al.*, 2007).

MECHANICAL-SEAL WATER MANAGEMENT.   Mechanical-seal water management is a device that helps to join systems or mechanisms together by stopping leakages by containing the water pressure and eliminating contamination, e.g. in plumbing systems. The mechanical seals can be used around pumps and other equipment promoting savings through better control of flush and quench water (Johnson, 2016).

FLOOR WASHERS.   Industrial floor-cleaning appliances are used in factories. Water-efficient scrubbers and driers (Zhang, 2017) are incorporated in wash-water recovery equipment, which receives the dirty water in the recovery tank and processes it so that it can be re-fed to the scrubber and drier solution tank for reuse (Venard and Bearup, 2016). These appliances can be used in combination with pressure-washing kits to the same end; however, collection of the dirty water is more cumbersome.

5.5.1.3. Reuse and recycling – technical solutions

RAINWATER HARVESTING.   Rainwater harvesting is the process of collecting rainwater, which would have drained into the drainage system or into open ground or been lost due to evaporation. The harvesting process can provide a good source of water in the factory for various non-drinking-water applications such as water flushing in toilets, use in washing machines, in cooling towers or for cleaning purposes (Envirowise, 2008). The Unilever factory in Durban, South Africa, harvested rainwater from its 22,000 m^2 roof area, which met almost 20% of the annual water demand (Water Scarcity Solutions, 2012).

COUNTER-CURRENT RINSING.   In comparison to other industries, the counter-current rinsing technique is scarcely applied in food factories. Rinse water can be more effectively utilized by moving a product through a series of stages and tanks (WRAP, 2005). At every stage, fresh or make-up, water is supplied; counter-current rinsing can be deployed as long as it complies with hygiene standards. In this system, the product is first rinsed using grey water and then gradually with cleaner water and at the same time the rinse water steadily moves from the clean water towards grey water. As per the WRAP (2005) report, using counter-current rinsing, a company managed to reduce their water consumption by 70%.

MEMBRANE SEPARATION.   Membrane separation uses a thin physical barrier through which materials can either filter through, get rejected or be retained. The nature of separation needed determines the structure and features of membranes. It has many applications within the food sector other than recovering water; it can also be utilized to concentrate and purify product and recover residues of raw materials and products from wastewater (Kotsanopoulos and Arvanitoyannis, 2015). Membrane filtration can be categorized into four main types, dependent on their pore sizes: microfiltration (MF), ultrafiltration (UF), nanofiltration (NF) and reverse osmosis (RO). Kowalik-Klimczak (2017) proposed a two-step system based on membrane filtration technique to treat dairies' wastewater. It further suggests using a combination of RO, UF or NF technique for regeneration of dairy wastewater.

VEHICLE WASHERS.   The equipment is built in such a way that it reduces the amount of water utilized in the washing process. This can be done through recycling the water used and by making changes to reduce the quantity of water used per washing (Rees *et al.*, 2003). For example, vehicle washing in red-meat slaughterhouses can use up to 5% of the overall water needed (AHDB, 2011). In order to reclaim or recycle the water, a trap is built beneath the vehicle to collect most or all of the water. The collected water is filtered and purified, usually by transferring to settlement tanks where larger particles are separated and the water goes through the refining process to reclaim the water fully (Mistry *et al.* 2007).

5.5.1.4. Effluent reduction and treatment – technical solutions

PIGGING.   Pigging is a technique for eradicating blockages in pipes, sewers and other equipment in the industry and is particularly suitable as the pipes need no dismantling for cleaning purposes (WRAP, 2013). It is regularly used in combination with clean-in-place systems and uses solid objects such as rubber or plastic plugs, or ice cubes or uses pressured air to remove any debris from the previous processes and cleans those to make ready for the next process. There are a number of benefits such as saving a large amount of water from cleaning pipes during changeover of products, product efficiencies through capturing final residues of the processes and incorporating it back into the final product (WRAP, 2013). If cleaning agents are used, then the amount needed for cleaning is reduced, which also reduces the wastewater load. In terms of pigging in the food industry, ice is the most flexible as it goes through pipework of various diameters and can become part of the finished food product.

SLURRY DE-WATERING/DRYING.   Slurry de-watering/drying is a technique to reduce the amount of water present in sludge and slurries by using centrifuging, pressing and high-temperature driers to squeeze out excess water for reuse. The resulting solid is called a cake; therefore, this process aims at reducing the weight of sludge and slurries and thereby reduces transportation costs (Hall. 2000; Kroiss and Zessner, 2007).

ELECTRO-COAGULATION.   Electro-coagulation is used for cleaning wastewater by making electricity pass through it in order to precipitate out dissolved and suspended materials (Abbas and Ali, 2018). The resulting water can be utilized in many applications such as cleaning activities and other on-site facilities. It reduces costs for clean water intake and wastewater discharge due to lesser volumes (Moussa *et al.*, 2017). When compared to chemical coagulation, electro-coagulation is cheaper as it does not require operational processes to add chemicals and apparently produces smaller volumes of sludge (Crini and Lichtfouse, 2019).

ONLINE ANALYSERS.   Online analysers are used to ensure that wastewater has attained the correct level of effluent load, often measured using total organic carbon as a primary indicator (Bourgeois *et al.*, 2001). They are important as they allow manufacturers to know if their wastewater discharge meets the consent levels, thus avoiding penalties. They also allow manufacturers to ensure that they are not overtreating the wastewater discharge and thus wasting money on energy costs. Online analysers can also be used to monitor the volumes of water used and help identify inefficiencies with regard to water usage (Cornelissen *et al.*, 2018). Other benefits of this technique are that they are instant and do not need any kind of sampling and testing, which often delays the process.

DOSING EQUIPMENT.   Optimizing the food ingredients' dosing quantities to correct levels ensures less wastage of materials and thereby less volume of water is needed for cleaning and washing utensils (Mistry *et al.*, 2007).

CLOSED-TRANSFER EQUIPMENT.   In this technique the products are moved from one vessel to another through pipework without the actual need for direct contact, ensuring no spillages between batch processes (Reichardt, 2018); thus, it avoids waste products being washed away to sewers and issues of cross-contamination.

ULTRA-VIOLET LIGHT, OZONE GAS AND IONIZING.   Ultra-violet light, ozone gas and ionizing radiation are used for treating wastewater to kill pathogens and oxidize trace chemicals making them easier to separate. It therefore minimizes the effluent load and makes it more suitable for reuse. Lee *et al.* (2015) compared the economical effectiveness of the three technologies in treating wastewater effluent.

SAND FILTERS.   In sand filter technique, layers of fine and coarse sands are used in conjunction with other materials in order to filter wastewater and physically remove suspended particles as well as dissolve material through adsorption. Bosak *et al.* (2016) examined sand filtration at a potato farm wastewater treatment system and discovered that it removed total soluble solids, biochemical oxygen demand, total phosphorus and total nitrogen at various levels. The filter needs to be periodically backwashed with clean water to achieve optimal performance and the resulting sludge needs further treatment or disposal.

DISSOLVED AIR FLOTATION.   Dissolved air flotation is used in conjunction with coagulation techniques to separate flocs and precipitates by passing gas through water in the upward direction. This forces the suspended material to the surface where it can be scraped off for disposal. For example, Radzuan *et al.* (2016) measured the oil droplet removal efficiency from oil-in-water mixtures using dissolved air flotation.

5.6. Chapter Summary

This chapter has reviewed the most significant initiatives taken to tackle food waste, energy and water issues in food manufacturing. Most of the techniques and technologies discussed are applied within food industries to minimize food waste and reduce energy and water consumption. The review has highlighted the lack of tools, techniques or technologies that can monitor and analyse food waste generation and energy and water consumption in real time to support such reduction in the food manufacturing sector.

References

Abbas, S.H. and Ali, W.H. (2018) Electrocoagulation technique used to treat wastewater: a review. *American Journal of Engineering Research* 7, 74–88.

Abdelaziz, E.A., Saidur, R. and Mekhilef, S. (2011) A review on energy saving strategies in the industrial sector. *Renewable and Sustainable Energy Reviews* 15, 150–168.

AHDB (2011) *Resource use in the British beef and lamb processing sector – water.* AHDB. Available at: http://beefandlamb.ahdb.org.uk/wp-content/uploads/2013/04/resourceusewater_100211-factsheet-2.pdf (accessed 12 February 2019).

Ahmed, I.I. and Gupta, A.K. (2010) Pyrolysis and gasification of food waste: syngas characteristics and char gasification kinetics. *Applied Energy* 87, 101–108.

Altmann, M., Michalski, J., Brenninkmeijer, A., Lanoix, J.-C., Tisserand, P. *et al.* (2010) *EU Energy Efficiency Policy – Achievements and Outlook.* European Parliament Committee on Industry, Research and Energy (ITRE). Available at: http://www.europarl.europa.eu/RegData/etudes/etudes/join/2010/451482/IPOL-ITRE_ET%282010%29451482_EN.pdf (accessed 1 April 2019].

Amazonas, M. (2009) Sustainable use of water in the food and beverage industry. *Asian Water*, September. Available at: http://www.thonhauser.net/files/5914/1404/3728/AW_Sept_Water_Footprint_150.pdf(accessed 25 April 2019).

Attwood, K. (2008) McCain's potato chip factory to be run on wind power. Available at: https://www.belfasttelegraph.co.uk/business/mccains-potato-chip-factory-to-be-run-on-wind-power-28060953.html(accessed 20 September 2021).

Arvanitoyannis, I.S. (2010) *Waste Management for the Food Industries.* Academic Press, Burlington, Massachusetts.

Arvanitoyannis, I.S., Ladas, D. and Mavromatis, A. (2006) Potential uses and applications of treated wine waste: a review. *International Journal of Food Science and Technology* 41, 475–487.

Barlow, C.Y. and Morgan, D.C. (2013) Polymer film packaging for food: an environmental assessment. *Resources, Conservation and Recycling* 78, 74–80.

Bavar, M., Sarrafzadeh, M.-H., Asgharnejad, H. and Horouzi-Firouz, H. (2018) Water management methods in food industry: corn refinery as a case study. *Journal of Food Engineering* 238, 78–84.

Best, R., Pilatowsky, I., Scoccia, R. and Motta, M. (2013) Solar cooling in the food industry in Mexico: a case study. *Applied Thermal Engineering* 50, 1447–1452.

Bilska, B., Wrzosek, M., Kołożyn-Krajewska, D. and Krajewski, K. (2016) Risk of food losses and potential of food recovery for social purposes. *Waste Management* 52, 269–277.

Blanke, M. and Burdick, B. (2005) Food (miles) for thought-energy balance for locally-grown versus imported apple fruit. *Environmental Science and Pollution Research* 12, 125–127.

Bosak, V.K., VanderZaak, A.C., Crolla, A., Kinsley, C., Chabot, D., Miller, S.S. and Gordon, R.J. (2016) Treatment of potato farm wastewater with sand filtration. *Environmental Technology* 37, 1597–1604.

Bourgeois, W., Burgess, J.E. and Stuetz, R.M. (2001) On-line monitoring of wastewater quality: a review. *Journal of Chemical Technology and Biotechnology* 76, 337–348.

Brunner, P.H. and Rechberger, H. (2015) Waste to energy – key elements for sustainable waste management. *Waste Management* 37, 3–12.

Buksti, M., Fremming, T., Juul, S., Grandjean, F. and Christensen, S. (2015) Surplus food redistribution system. European Community Seventh Framework Programme. Available at: https://www.eu-fusions.org/phocadownload/feasibility-studies/SMS/Fusions%20Surplus%20Food.pdf (accessed 6 January 2019).

Burfoot, D., Reavell, S., Wilkinson, D. and Duke, N. (2004) Localised air delivery to reduce energy use in the food industry. *Journal of Food Engineering* 62, 23–28.

Carbon Trust (2018) Better business guide to energy saving. Carbon Trust. Available at: https://www.carbontrust.com/resources/guides/energy-efficiency/better-business-guide-to-energy-saving/ (accessed 27 January 2019).

Caton, P.A., Carr, M.A., Kim, S.S. and Beautyman, M.J. (2010) Energy recovery from waste food by combustion or gasification with the potential for regenerative dehydration: a case study. *Energy Conversion and Management* 51, 1157–1169.

Chiu, S.L. and Lo, I.M. (2016) Reviewing the anaerobic digestion and co-digestion process of food waste from the perspectives on biogas production performance and environmental impacts. *Environmental Science and Pollution Research* 23, 24435–24450.

Cornelissen, R., Van Dyck, T., Dries, J., Ockier, P., Smets, I. *et al.* (2018) Application of online instrumentation in industrial wastewater treatment plants – a survey in Flanders, Belgium. *Water Science and Technology* 78, 957–967.

Crini, G. and Lichtfouse, E. (2019) Advantages and disadvantages of techniques used for wastewater treatment. *Environmental Chemistry Letters* 17, 145–155.

Damour, C., Hamdi, M., Josset, C., Auvity, B. and Bollereaux, L. (2012) Energy analysis and optimization of a food defrosting system. *Energy* 37, 562–570.

Darlington, R. and Rahimifard, S. (2006) A responsive demand management framework for the minimization of waste in convenience food manufacture. *International Journal of Computer Integrated Manufacturing* 19, 751–761.

David, E.L. (1990) Trends and associated factors in offstream water use manufacturing and mining water use in the United States, 1954–83. *National Water Summary* 2775, 81.

DEFRA (2013) *Incineration of Municipal Solid Waste.* Department for Environment, Food and Rural Affairs. Available at: https://assets.publishing.service.gov.uk/government/uploads/system/uploads/attachment_data/file/221036/pb13889-incineration-municipal-waste.pdf (accessed 13 March 2019).

DEFRA (n.d.) *Feeding Catering Waste to Farmed Animals is ILLEGAL.* Department for Environment, Food and Rural Affairs. Available at: https://www.chippingnortonvets.co.uk/files/DEFRA%20catering%20waste%20leaflet.pdf (accessed 12 February 2019).

EIA (2017) International Energy Outlook 2017. US Energy Information Administration. Available at: https://www.eia.gov/outlooks/ieo/pdf/0484(2017).pdf (accessed 23 March 2019).

Eichhammer, W. *et al.* (2009) *Study on the energy savings potentials in EU member states, candidate countries and EEA countries: final report for the European Commission Directorate-General Energy and Transport.* Available at: https://ec.europa.eu/energy/sites/ener/files/documents/2009_03_15_esd_efficiency_potentials_final_report.pdf (accessed 20 September 2021).

Ellis, M., Dillich, S. and Margolis, N. (2001) *Industrial Water Use and Its Energy Implications.* US Deptartment of Energy, Office of Energy Efficiency and Renewable Energy, Washington, DC.

Emkes, H., Coulon, F. and Wagland, S. (2015) A decision support tool for landfill methane generation and gas collection. *Waste Management* 43, 307–318.

Environment Agency (2010) *Environmental Permitting Regulations (England and Wales) 2010 – Understanding the Landfill Directive.* Environment Agency. Available at: https://assets.publishing.service.gov.uk/government/uploads/system/uploads/attachment_data/file/296536/LIT_8286_f89fa7.pdf (accessed 23 March 2019).

Environment Agency (2013) *How to Comply with Your Landspreading Permit.* Environment Agency. Available at: https://assets.publishing.service.gov.uk/government/uploads/system/uploads/attachment_data/file/290130/LIT_5492_40c081.pdf (accessed 13 March 2019).

Environment Protection Agency (2004) *Landspreading of Organic Waste – Guidance on Groundwater Vulnerability Assessment of Land.* Environment Protection Agency Johnstown Castle, Ireland.

Envirowise (2008) Reducing mains water use through rainwater harvesting – collecting a valuable free resource and reducing the impact of runoff. WRAP.

FAO (2013) *Food Wastage Footprint: Impacts on Natural Resources – Technical Report.* Food and Agriculture Organization of the United Nations, Natural Resources and Management Department. Available at: http://www.fao.org/3/ar429e/ar429e.pdf (accessed 17 January 2019).

FAO (2017) *The Future of Food and Agriculture – Trends and Challenges.* Food and Agriculture Organization of the United Nations. Available at: http://www.fao.org/3/a-i6583e.pdf (accessed 21 December 2020).

Fava, F., Zanaroli, G., Vannini, L., Guerzoni, E., Bordoni, A. *et al.* (2013) New advances in the integrated management of food processing by-products in Europe: sustainable exploitation of fruit and cereal processing by-products with the production of new food products (NAMASTE EU). *New Biotechnology* 30, 647–655.

Fisher, K., James, K., Sheane, R., Nippress, J., Allen, S.R. *et al.* (2013) An initial assessment of the environmental impact of grocery products. Available at: http://www.wrap.org.uk/sites/files/wrap/An%20initial%20assessment%20of%20the%20environmental%20impact%20of%20grocery%20products%20final_0.pdf (accessed 15 January 2019).

FoodDrinkEurope (2014) Preventing food wastage in the food and drink sector – Europe's food and drink manufacturers take action to prevent food wastage. FoodDrinkEurope. Available at: https://www.fooddrinkeurope.eu/uploads/publications_documents/Preventing_food_wastage_in_the_food_and_drink_sector.pdf (accessed 17 January 2019).

Galanakis, C.M. (2015) *Food Waste Recovery: Processing Technologies and Industrial Techniques.* Academic Press, London.

Girotto, F., Alibardi, L. and Cossu, R. (2015) Food waste generation and industrial uses: a review. *Waste Management* 45, 32–41.

Giuseppe, A., Mario, E. and Cinzia, M. (2014) Economic benefits from food recovery at the retail stage: an application to Italian food chains. *Waste Management* 34, 1306–1316.

Gould, J., Garcia-Garcia, G. and Wolf, B. (2016) Pickering particles prepared from food waste. *Materials* 9, 791.

Gov.uk (2014) *Guidance – How Food Businesses Must Dispose of Food and Former Foodstuffs.* Department for Environment, Food and Rural Affairs and Animal and Plant Health Agency. Available at: https://www.gov.uk/guidance/how-food-businesses-must-dispose-of-food-and-former-foodstuffs (accessed 13 March 2019).

Grayson, D. and Hodges, A. (2017) *Corporate Social Opportunity! Seven Steps to Make Corporate Social Responsibility Work for Your Business.* Routledge, Abingdon, UK.

Griffin, P.W., Hammond, G.P. and Norman, J.B. (2016) Industrial energy use and carbon emissions reduction: a UK perspective. *Wiley Interdisciplinary Reviews: Energy and Environment* 5, 684–714.

Guendehou, G.H.S., Koch, M., Hockstad, L., Pipatti, R. and Yamada, M. (2006) *Incineration and Open Burning of Waste. 2006 IPCC Guidelines for National Greenhouse Gas Inventories.* Intergovernmental Panel on Climate Change, Geneva.

Hall, G.M. and Howe, J. (2012) Energy from waste and the food processing industry. *Process Safety and Environmental Protection* 90, 203–212.

Hall, J. (2000) *Ecological and Economical Balance for Sludge Management Options.* European Communities, Stresa, Italy.

Henningsson, S., Hyde, K., Smith, A. and Campbell, M. (2004) The value of resource efficiency in the food industry: a waste minimisation project in East Anglia, UK. *Journal of Cleaner Production* 12, 505–512.

Hoornweg, D. and Bhada-Tata, P. (2012) *What a Waste: A Global Review of Solid Waste Management.* World Bank, Washington, DC.

IEA (2018) *World Energy Outlook 2018.* International Energy Agency. Available at: https://www.iea.org/reports/world-energy-outlook-2018 (accessed 21 December 2020).

IGD (2007) Water use in the supply chain. The Institute of Grocery Distribution. Available at: https://www.igd.com/articles/article-viewer/t/water-use-in-the-supply-chain/i/15519 (accessed 17 September 2018).

IRENA (2015) Renewable energy options for the industry sector: global and regional potential until 2030. International Renewable Energy Agency. Available at: https://irena.org/media/Files/IRENA/Agency/Articles/2016/Nov/IRENA_RE_Potential_for_Industry_BP_2015.pdf?la=en&hash=1214D8FDBD507297FC61073DACE78F8F31927663 (accessed 15 April 2019).

Jagtap, S. (2019) Utilising the Internet of Things concepts to improve the resource efficiency of food manufacturing. PhD thesis, Loughborough University, UK.

Jagtap, S. and Rahimifard, S. (2017) Utilisation of Internet of Things to improve resource efficiency of food supply chains. In: Salampasis, M., Theodoridis, A. and Bournaris, T. (eds) *Proceedings of the 8th International Conference on Information and Communication Technologies in Agriculture, Food and Environment (HAICTA 2017).* Chania, Greece, 21–24 September. CEUR Workshop Proceedings, vol. 2030, pp. 8–19.

Jagtap, S. and Rahimifard, S. (2018) Real-time data collection to improve energy efficiency in food manufacturing. International Congress on Organizational Management, Energy Efficiency and Occupational Health and Safety in Agrifood Industry. CEi, Castelo Branco, Portugal, 3–4 October.

Jagtap, S. and Rahimifard, S. (2019) Unlocking the potential of the Internet of Things to improve resource efficiency in food supply chains. In: *International Conference on Information and Communication Technologies in Agriculture, Food & Environment.* Springer, Cham, Switzerland, pp. 287–301.

Jagtap, S., Skouteris, G., Choudhari, V. and Rahimifard, S. (2018) *Improving Water Efficiency in the Beverage Industry with the Internet of Things.* Changa, Anand, s.n.

Jagtap, S., Rahimifard, S. and Duong, L.N.K. (2019) Real-time data collection to improve energy efficiency: a case study of food manufacturer. *Journal of Food Processing and Preservation,* e14338.

Jensen, J.K. and Govindan, K. (2014) Assessment of renewable bioenergy application: a case in the food supply chain industry. *Journal of Cleaner Production* 66, 254–263.

Johnson, D. (2016) APE pumps. *Water & Sanitation Africa* 11, 35.

Kamiński, J. and Leduc, G. (2010) Energy efficiency improvement options for the EU food industry. *Polityka Energetyczna* 13, 87–97.

Kannan, R. and Boie, W. (2003) Energy management practices in SME: case study of a bakery in Germany. *Energy Conversion and Management* 44, 945–959.

Kirby, R.M., Bartram, J. and Carr, R. (2003) Water in food production and processing: quantity and quality concerns. *Food Control* 14, 283–299.

Kotsanopoulos, K.V. and Arvanitoyannis, I.S. (2015) Membrane processing technology in the food industry: food processing, wastewater treatment, and effects on physical, microbiological, organoleptic, and nutritional properties of foods. *Critical Reviews in Food Science and Nutrition* 55, 1147–1175.

Kowalik-Klimczak, A. (2017) The possibilities of using membrane filtration in the dairy industry. *Journal of Machine Construction and Maintenance-Problemy Eksploatacji* 105, 99–108.

Krasulya, O., Shestakov, S., Bogush, V. and Irina, P. (2014) Applications of sonochemistry in Russian food processing industry. *Ultrasonics Sonochemistry* 21, 2112–2116.

Kroiss, H. and Zessner, M. (2007) *Ecological and Economical Relevance of Sludge Treatment and Disposal Options*. Available at: chrome-extension://oemmndcbldboiebfnladdacbdf-madadm/http://citeseerx.ist.psu.edu/viewdoc/download?doi=10.1.1.462.7993&rep=rep1&type=pdf (accessed 13 September 2021).

Kwak, T.-H., Maken, S., Lee, S., Park, J.-W., Min, B. and Yoo, Y.D. (2006) Environmental aspects of gasification of Korean municipal solid waste in a pilot plant. *Fuel* 85(14-15), 2012–2017.

Lee, O.M., Kim, H.Y., Park, W., Kim, T.-H. and Yu, S. (2015) A comparative study of disinfection efficiency and regrowth control of microorganism in secondary wastewater effluent using UV, ozone, and ionizing irradiation process. *Journal of Hazardous Materials* 295, 201–208.

Lin, C.S.K., Pfaltzgraff, L.A., Herrero-Davila, L., Mubofu, E.B., Abderrahim, S. *et al.* (2013) Food waste as a valuable resource for the production of chemicals, materials and fuels. *Current situation and global perspective. Energy & Environmental Science* 6, 426–464.

Lin, C.S.K., Koutinas, A.A., Stamatelatou, K., Mubofu, E.B., Matharu, A.S. *et al.* (2014) Current and future trends in food waste valorization for the production of chemicals, materials and fuels: a global perspective. *Biofuels, Bioproducts and Biorefining* 8, 686–715.

Lohri, C.R., Sweeney, D. and Rajabu, H.M. (2015) Carbonizing urban biowaste for low-cost char production in developing countries. A review of knowledge, practices and technologies. Eawag. Available at: https://www.eawag.ch/fileadmin/Domain1/Abteilungen/sandec/E-Learning/Moocs/Solid_Waste/W4/Carbonazing_urban_biowaste_low_cost_char_2015.pdf (accessed 23 February 2019).

Lu, B.-S., Lee, M., Chen, S.-T., Chen, C.-H., Luo, Y.-C. and Den, W. (2018) Strategic optimization of water reuse in wafer fabs via multi-constraint linear programming technique. *Water-Energy Nexus* 1, 86–96.

Matharu, A.S., de Melo, E.M. and Houghton, J.A. (2016) Opportunity for high value-added chemicals from food supply chain wastes. *Bioresource Technology* 215, 123–130.

Meeker, D.L. (2006) Essential rendering – all about the animal by-products industry. National Renderers Association. Available at: http://assets.nationalrenderers.org/essential_rendering_book.pdf (accessed 13 March 2019).

Mekhilef, S., Saidur, R. and Safari, A. (2011) A review on solar energy use in industries. *Renewable and Sustainable Energy Reviews* 15, 1777–1790.

Miah, J.H., Griffiths, A., McNeil, R., Poonaji, I., Martin, R., Yang, A. and Morse, S. (2014) Heat integration in processes with diverse production lines: a comprehensive framework and an application in food industry. *Applied Energy* 132, 452–464.

Mistry, P., Cadman, J., Miller, S., Ogilvie, S. and Pugh, M. (2007) Resource use efficiency in food chains: priorities for water, energy and waste opportunities. AEA Energy & Environment. Available at: sciencesearch.defra.gov.uk/Document.aspx?Document=WU0103_4830_FRA.pdf (accessed 11 March 2019).

Monforti-Ferrario, F., Dallemand, J.-F., Pinedo Pascua, I., Motola, V., Banja, M. *et al.* (2015) *Energy Use in the EU Food Sector: State of Play and Opportunities for Improvement*. European Commission, Joint Research Centre, Institute for Energy and Transport and Institute for Environment and Sustainability. Available at: http://publications.jrc.ec.europa.eu/repository/bitstream/JRC96121/ldna27247enn.pdf (accessed 11 February 2019).

Morawicki, R.O. (2012) *Handbook of Sustainability for the Food Sciences.* Wiley-Blackwell, Chichester, UK.

Moussa, D.T., El-Naas, M.H., Nasser, M. and Al-Marri, M.J. (2017) A comprehensive review of electrocoagulation for water treatment: potentials and challenges. *Journal of Environmental Management* 186, 24–41.

Muller, D.C., Marechal, F.M., Wolewinski, T. and Roux, P.J. (2007) An energy management method for the food industry. *Applied Thermal Engineering* 27, 2677–2686.

Palabiyik, I., Yilmaz, M.T., Fryer, P.J., Robbins, P.T. and Toker, O.S. (2015) Minimising the environmental footprint of industrial-scaled cleaning processes by optimisation of a novel clean-in-place system protocol. *Journal of Cleaner Production* 108, 1009–1018.

Parfitt, J., Stanley, C. and Thompson, L. (2016) *Guidance for Food and Drink Manufacturers and Retailers on the Use of Food Surplus as Animal Feed.* Available at: https://wrap.org.uk/sites/default/files/2020-09/WRAP-2016-05-17-Animal-Feed-Guidance-v1.0-for-publication.pdf (accessed 20 September 2021).

Pathare, P.B., Roskilly, A.P. and Jagtap, S. (2019) Energy efficiency in meat processing. In: Gaspar, D.G. and da Silva, P.D. (eds) *Novel Technologies and Systems for Food Preservation.* IGI Global, Hershey, Pennsylvania, pp. 78–107.

Paudel, E., Van der Sman, R.G.M., Westerik, N., Ashutosh, A., Dewi, B. and Boom, M. (2017) More efficient mushroom canning through pinch and exergy analysis. *Journal of Food Engineering* 195, 105–113.

Peelman, N., Ragaert, P., De Meulnaer, B., Adons, D., Peeters, R. *et al.* (2013) Application of bioplastics for food packaging. *Trends in Food Science & Technology* 32, 128–141.

Pirasteh, G., Saidur, R., Rahman, S.M.A. and Rahim, N.A. (2014) A review on development of solar drying applications. *Renewable and Sustainable Energy Reviews* 31, 133–148.

Pleissner, D., Qi, Q., Gao, C., Rivero, C.P., Webb, C., Lin, C.S.K. and Venus, J. (2016) Valorization of organic residues for the production of added value chemicals: a contribution to the bio-based economy. *Biochemical Engineering Journal* 116, 3–16.

Ptasinski, K.J., Koymans, M.N. and Verspagen, H.H.G. (2006) Performance of the Dutch energy sector based on energy, exergy and extended exergy accounting. *Energy* 31, 3135–3144.

Puligundla, P., Jung, J. and Ko, S. (2012) Carbon dioxide sensors for intelligent food packaging applications. *Food Control* 25, 328–333.

Radzuan, M.A., Belope, M.A.B. and Thorpe, R.B. (2016) Removal of fine oil droplets from oil-in-water mixtures by dissolved air flotation. *Chemical Engineering Research and Design* 115, 19–33.

Rahimifard, S., Woolley, E., Webb, D.P., Garcia-Garcia, G., Stone, J. *et al.* (2017) Forging new frontiers in sustainable food manufacturing. In: *International Conference on Sustainable Design and Manufacturing.* Springer, Cham, Switzerland, pp. 13–24.

Ravindran, R. and Jaiswal, A.K. (2016) Exploitation of food industry waste for high-value products. *Trends in Biotechnology* 34, pp. 58–69.

Rees, B, Cessford, F., Connelly, R., Cowan, J. and Bowell, R. (2003) *Optimum Use of Water for Industry and Agriculture: Phase 3.* Environment Agency, Bristol, UK.

Reichardt, K. (2018) How industrial facilities can transfer from water use to water efficiency. Environmental protection. Available at: https://eponline.com/articles/2018/05/10/how-industrial-facilities-transfer-water-efficiency.aspx (accessed 2 January 2019).

Reynolds, C., Piantadosi, J. and Boland, J. (2015) Rescuing food from the organics waste stream to feed the food insecure: an economic and environmental assessment of Australian food rescue operations using environmentally extended waste input-output analysis. *Sustainability* 7, 4707–4726.

Romero-Güiza, M.S., Vila, J., Mata-Alvarez, J., Chimenos, J.M. and Astals, S. (2016) The role of additives on anaerobic digestion: a review. *Renewable and Sustainable Energy Reviews* 58, 1486–1499.

Russell, D.A. (2014) Sustainable (food) packaging: an overview. *Food Additives & Contaminants: Part A* 31, 396–401.

Salemdeeb, R., zu Ermgassen, E.K.H.J., Kim, M.H., Balmford, A. and Al-Tabbaa, A. (2017) Environmental and health impacts of using food waste as animal feed: a comparative analysis of food waste management options. *Journal of Cleaner Production* 140, 871–880.

San Martin, D., Ramos, S. and Zufía, J. (2016) Valorisation of food waste to produce new raw materials for animal feed. *Food Chemistry* 198, 68–74.

Schaub, S.M. and Leonard, J.J. (1996) Composting: an alternative waste management option for food processing industries. *Trends in Food Science & Technology* 7, 263–268.

Schlich, E. and Fleissner, U. (2005) The ecology of scale: assessment of regional energy turnover and comparison with global food. *The International Journal of Life Cycle Assessment* 10, 219–223.

Seck, G.S., Guerassimoff, G. and Maïzi, N. (2013) Heat recovery with heat pumps in non-energy intensive industry: a detailed bottom-up model analysis in the French food & drink industry. *Applied Energy* 111, 489–504.

Sen, B., Aravind, J., Kanmani, P. and Lay, C.H. (2016) State of the art and future concept of food waste fermentation to bioenergy. *Renewable and Sustainable Energy Reviews* 53, 547–557.

Stuart, T. (2009) *Waste: Uncovering the Global Food Scandal.* W.W. Norton, New York.

Tassou, S.A., De-Lille, G. and Ge, Y.T. (2009) Food transport refrigeration: approaches to reduce energy consumption and environmental impacts of road transport. *Applied Thermal Engineering* 29, 1467–1477.

Tassou, S.A., Lewis, J.S., Ge, Y.T., Hadawey, A. and Chaer, I. (2010) A review of emerging technologies for food refrigeration applications. *Applied Thermal Engineering* 30, 263–276.

Tassou, S.A., Ge, Y., Hadawey, A. and Marriott, D. (2011) Energy consumption and conservation in food retailing. *Applied Thermal Engineering* 31, 147–156.

Thi, N.B.D., Kumar, G. and Lin, C.Y. (2015) An overview of food waste management in developing countries: current status and future perspective. *Journal of Environmental Management* 157, 220–229.

Thollander, P. and Palm, J. (2013) Managing energy efficiency in industry. In: Thollander, P. and Palm, J. (authors) *Improving Energy Efficiency in Industrial Energy Systems.* London: Springer, pp. 85–104.

UNEP and ISWA (2015) Global waste management outlook. United Nations Environment Programme. Available at: https://www.uncclearn.org/sites/default/files/inventory/unep23092015.pdf (accessed 12 March 2019).

United Nations (n.d.) *Water, Food and Energy.* Available at: https://www.unwater.org/water-facts/water-food-and-energy/ (accessed 1 September 2020).

Vallack, H., Timmis, A., Robinson, K., Sato, M., Kroon, P. and Plomp, A. (2011) Technology innovation for energy intensive industry in the United Kingdom. The Centre for Low Carbon Futures. Available at: https://www.researchgate.net/publication/277189547_Technology_Innovation_for_Energy_Intensive_Industry_in_the_United_Kingdom (accessed 29 July 2021).

Venard, D.C. and Bearup, A. (2016) Floor cleaning tool having a mechanically operated pump. United States, Patent No. US9301661B2.

Waldron, K. (2007) Waste minimization, management and co-product recovery in food processing: an introduction. In: Waldron, K. (ed.) *Handbook of Waste Management and Co-product Recovery in Food Processing,* Volume 1. Woodhead Publishing, Oxford, UK, pp. 3–20.

Water Scarcity Solutions (2012) *Water Recycling in the Food Sector.* Water Scarcity Solutions. Durban, South Africa. Available at: https://www.waterscarcitysolutions.org/wp-content/uploads/2015/08/Water-recycling-in-the-food-sector-Durban-South-Africa.pdf (accessed 17 January 2019).

Worrell, E., Bernstein, L., Roy, J., Price, L. and Harnisch, J. (2009) Industrial energy efficiency and climate change mitigation. *Energy Efficiency* 2, 109.

WRAP (2005) Cost-effective water saving devices and practices – for industrial sites. Available at: http://www.wrap.org.uk/sites/files/wrap/GG523_industrial%20Cost-effective%20 water%20saving%20devices%20and%20practices%20-%20for%20industrial%20sites. pdf (accessed 15 January 2019).

WRAP (2014) *Benefits of water metering and monitoring for the food & drink and hospitality & food service sectors.* Available at: http://www.wrap.org.uk/sites/files/wrap/WRAP%20 Metering%20and%20monitoring%20June%202014.pdf (accessed 15 January 2019).

WRAP (2017) Estimates of food surplus and waste arisings in the UK. Available at: https:// wrap.org.uk/resources/report/estimates-food-surplus-and-waste-arisings-uk-2017 (accessed 11 January 2019).

Xu, F., Li, Y., Ge, X., Yang, L. and Li, Y. (2018) Anaerobic digestion of food waste – challenges and opportunities. *Bioresource Technology* 247, 1047–1058.

Zhang, X. (2017) Floor Washer Cleaning Device and Floor Washer. United States Patent No. US9844311B2.

6 Sustainability in the Food Supply Chains

LINH DUONG

6.1. Introduction

Nowadays, thanks to globalization, we can easily have New Zealand lamb or Canadian lobster, no matter where we are. For pasta eaten in the UK, most is made from wheat shipped from Canada to processing companies in Brazil or Italy. After a long journey through Europe, UK distributors sell it to supermarkets (Terazono and Evans, 2020). These examples represent the globalization characteristic of our world's economy, which has revolutionized food supply chains in recent years. As a result, it increases food availability around the globe. However, it has also led to a situation where many countries are dependent, partially or heavily, on imported food. This globalization has created opportunities and risks for the food supply chains. The purpose of this chapter is to consider global food supply chains and how they evolve to achieve one of the biggest missions in this century – sustainable development.

6.2. Sustainability

Sustainable development or sustainability is defined as 'the development that meets the needs of the present without compromising the ability of future generations to meet their own needs' (WCED, 1987). In some instances, sustainability and green issues are interchangeable. However, we believe that the green issue is just one dimension of sustainability. This chapter considers three dimensions or the triple-bottom-line of sustainability, including environmental, economic and social issues. These dimensions are also referred to as 3Ps – people, profit and planet. The triple-bottom-line offers a framework to measure the performance of a business. It reminds us that to be sustainable a business must keep a balanced focus on economic, environmental and social impacts when undertaking any activity.

© CAB International 2022. *Food Industry 4.0: Unlocking Advancement Opportunities in the FoodManufacturing Sector* (W. Martindale *et al.*)
DOI: 10.1079/9781789248593.0006

In the triple-bottom-line, the economic impact often relates to issues such as the profitability of firms, financial security, or people's incomes. It focuses on the economic contribution of a business to the growth of the surrounding system. Environmental impact refers to the use of resources and the footprint that a firm generates during its operations. It often relates to issues such as climate change, flooding or the depletion of resources. Finally, social impact focuses on the value for both employees and the surrounding community. It means that a firm should provide equal opportunities, ensure quality of life, or encourage diversity, and often relates to poverty, working and living conditions, or gender equality.

The triple-bottom-line concept has been adopted widely by numerous agencies and organizations. For example, the UN highlighted that the achievement of sustainability should be based on balancing environmental, social and economic demands. They stated that '[w]e reaffirm that development is a central goal in itself and that sustainable development in its economic, social and environmental aspects constitutes a key element of the overarching framework of United Nations activities' (United Nations, 2005).

6.3. Sustainable Food Supply Chain

On the other side, supply chain management is defined as:

> the systemic, strategic coordination of the traditional business functions and the tactics across these business functions within a particular company and across businesses within the supply chain, for the purposes of improving the long-term performance of the individual companies and the supply chain as a whole.
>
> (Mentzer *et al.*, 2001, p. 18)

Supply chain management includes the management or exchange of materials and information from purchasing raw materials to delivering products or services to end-users. It aims at maximizing the overall value to all partners involved (e.g. suppliers, manufacturers and customers) (Lee *et al.* 1997). Thus, its abilities in responding to the three issues of environment, society and economy are the core concepts of a sustainable supply chain (Chardine-Baumann and Botta-Genoulaz 2014).

A sustainable supply chain concerns:

> [the] management of material, information and capital flows as well as cooperation among companies along the supply chain while taking goals from all three dimensions of sustainable development, i.e., economic, environmental and social, into account which are derived from customer and stakeholder requirements.
>
> (Seuring and Müller, 2008, p. 1700)

Focusing on a sustainable supply chain is essential for any business for two reasons. First, a company has to consider the energy and other resources it uses and the footprint it generates. It is noted that manufacturing, transportation and new product development contribute most to the footprint; thus, supply chain management has a vital role in reducing the footprint (Kleindorfer *et al.*,

2005). Second, companies have to operate in a responsive way and consider the quality of life of employees and the external community (Gimenez *et al.*, 2012). As a supply chain involves many partners, it can significantly affect any effort towards sustainability.

More recently, executives of food companies are aware that the success of their supply chains will determine their competitive advantage. Food supply chains have received much interest, and food products are shipped across regions or countries with a wide range of partners involving processing, storage, transportation and distribution operations. Sustainable food supply chains refer to all processes in the food chain, from procurement to production and distribution. They also include the reverse processes like the collection of unused food products and ensuring social, economic and environmental values (Bloemhof *et al.*, 2015). The increased awareness and consumer demands on sustainability have put much pressure on the food supply chains. Firms can implement sustainability programmes such as waste reduction, temperature-controlled distribution, and high-standard foodservice operations (Mangla *et al.*, 2019). The concept of sustainable food supply chains has gained significant attention in recent years (Kamble *et al.*, 2020).

6.3.1. Characteristics of food supply chains

Food supply chains have many unique characteristics that differentiate them from other product supply chains. The fundamental difference is that the quality of food products changes significantly and continuously throughout the entire supply chain (Duong *et al.*, 2020a). Additionally, food supply chains cover a wide range of products from fresh (e.g. flowers, vegetables) to processed (e.g. ready-to-eat meals, snacks). They also involve many partners such as growers, wholesalers, importers, exporters, transporters and retailers to bring products to consumers. Terms such as 'grass-to-glass' represent the whole process from the cows' grass to the dairy products. 'Farm-to-fork' represents the processing of raw materials from the farm into food for consumer demand. They symbolize the complexity of food supply chains. Investments in the food supply chain should aim to preserve food quality and improve the performance of the whole chain (Bloemhof *et al.*, 2015).

Box 6.1. The coffee supply chain. (From Wallach, 2020)

The Coffee Supply Chain: From Bean to Brew

Coffee is probably one of the most traded food items in the world and reveals the long and complex supply chain. It is estimated that, globally, over 2.25 billion cups of coffee are consumed every day. We can have a cup of coffee at any time of the day and at any place. However, there are only a few countries that grow the coffee plants. The coffee plants grow in humid climates, making Asia, South America and Africa optimal locations for production. After the coffee beans are dried out, they are shipped around the world through a global supply network. Then the coffee beans are roasted, and the final steps of the supply chain (i.e. grinding, brewing and drinking) can happen anywhere.

6.4. Issues and Challenges in Sustainable Food Supply Chains

Although there is an increasing interest from governments and other related partners, many gaps remain in the sustainable food supply chain (Jabbour et al., 2019). From the environmental perspective, the use of natural resources like water (Maier et al., 2020), the use of chemical fertilizers and pesticides (Taghikhah et al., 2021), or the effects of climate change and greenhouse gas emissions from food production (Khanal et al., 2019) are some major issues affecting the food supply chain. From the economic perspective, food contamination, food waste or operational inefficiencies and food counterfeiting affect the profitability of the food industry (Reynolds and Dolasinski, 2019; Nardi et al., 2020). Finally, food insecurity and consumers' diet habits are commonly cited social issues of food supply chains (Sibhatu and Qaim, 2018; Patra et al., 2020). This section highlights some major issues and challenges that affect a sustainable food supply chain's environmental, economic and social dimensions.

6.4.1. Food price volatility

The last decade has witnessed rapid changes in the global food price (Assefa et al., 2015). Demand factors such as economic growth and food habits, and supply factors such as climate change and the lack of logistics, are some antecedents for the food price volatility. These factors can cause a sudden increase in demand or decrease in supply, leading to an imbalance between demand and supply and creating an unpredictable change in food prices.

Food price volatility has impacts on all partners along the food supply chain. At the end of the chain, a small increase in the food price can strongly influence food security, especially for individuals in low- and middle-income groups, who spend most of their income on food (Béné, 2020). Farmers might reduce investments and output supply to cope with uncertainty of the food price (Wossen et al., 2018). Food price volatility leads to low operational efficiencies for food companies as they need to seek alternative sources and increase the inventory level (Taghikhah et al., 2021). These wide implications of food price volatility affect all three dimensions of sustainability for all partners in the food chain.

Box 6.2. (From Terazono, 2020)

Food Price Sparks Put Pressure on Developing Countries

Stockpiling, logistical bottlenecks, and dry weather have pushed wheat, soybeans, rice, and corn markets higher. According to the UN, the Food and Agriculture Organization's food price index hit a six-year high in November. 'The real impact is the access to food. People have lost their income. There are a lot of unhappy people and this is a recipe for social unrest,' said Abdolreza Abbassian, Senior Economist at the FAO.

This causes more problems for poor countries, particularly in Asia where the rebound economy after Covid-19 led to higher demand for soybeans and grains.

6.4.2. Food quality and safety

Food quality and safety management have high impacts on sustainability because food products require a specific range of temperature and humidity to maintain their quality (van der Vorst *et al.*, 2009). Moreover, food quality and safety incidents, including high-profile incidents of melamine contamination (Kong *et al.*, 2019), can lead to an expensive food recall campaign. Such emerging issues call for the adoption of quality and safety management systems. For example, Hazard Analysis Critical Control Point (HACCP), Good Manufacturing Practice (GMP) and Good Agriculture Practice (GAP) are common systems in many countries worldwide.

Due to globalization and international trade policies, the long and complex food supply chains put more pressure on managing food origin, food quality, food nutrition and ethical issues to achieve sustainability. Consumers are now more sensitive to the origin and processing of food products. For example, there is a huge debate about avoiding hormone-treated beef and chlorinated chicken to support British farmers (Jones, 2020). Additionally, the use of child labour or the wrong behaviour of suppliers/manufacturers poses a serious concern for sustainability. These issues show that food quality and safety can happen at any place in the food chain – farmers, middlemen, transporters, suppliers or manufacturers. Hence, it is important to build relationships and trust among partners within the food supply chain (Liu *et al.*, 2019). It is also more important to have management systems that can trace, track and increase information sharing along the food chain. In this instance, digital technologies, blockchain technologies, or robotics and autonomous systems are being advocated for use in the food chain (Cole *et al.*, 2019; Duong *et al.*, 2020a).

6.4.3. Food waste and loss

Another concern for sustainability is food waste and loss, as nearly one third of food products is wasted or lost (FAO, 2017). From the supply chain management perspective, food loss decreases the quality and quantity of food that occurs from post-harvest up to, but not including, the retail level. In contrast, food waste decreases the quality and quantity of food at the retail and consumer levels (FAO, 2017). According to this definition, food loss may refer to the disposed or discarded food at farmers or manufacturers; food waste may refer to the discarded food at household kitchens. It is worth noting that post-harvest waste and loss happen less in developed countries than in undeveloped countries, maybe because of the high-level adoption of robotics and autonomous systems in developed countries.

Food waste and loss cause significant concerns for the environment (e.g. water and energy to produce the unused food), economy (e.g. loss of profit), and society (e.g. food insecurity). Many models have been developed to reduce or eliminate food waste and loss (e.g. Bustos and Moors, 2018; Schanes *et al.*, 2018). Pohlmann *et al.* (2020) indicated a strategic role of the focal company in the vertical integration of the poultry supply chain, which aims to promote radical changes in business models and addresses the waste issue. Dora *et al.* (2020) tried to map food loss in the food processing industry. Based on a sample of 47 food processing companies in Belgium, the authors showed that food loss occurs mainly in the processing stage. They are just some recent efforts in preventing food waste and loss. In general, researchers have paid much attention to mapping food waste and loss, redesigning business models, developing performance measures, upgrading facilities (e.g. digitization), or optimizing the operations (e.g. optimizing the forecasting process).

6.5. How to Measure Sustainability in Food Supply Chains

Previous sections highlighted major issues in a food supply chain and why a sustainable food supply chain is critical. This section discusses methods to measure or ensure that a food supply chain is sustainable. To help decision-makers measure or assess the sustainability of the food supply chain, we need comprehensive performance measurement systems that cover all three sustainability dimensions. In an inventory management model, traditional financial measures have been criticized as too one-dimensional (Duong *et al.*, 2020a); thus, researchers (e.g. Duong *et al.*, 2018) advocate the use of non-financial measures. Similarly, many efforts have been devoted to developing measures to assess a sustainable food supply chain's economic and environmental and social aspects. Performance measures for sustainability are essential as they indicate the process towards sustainability. They also help to benchmark with competitors; finally, they support the decision-makers.

Despite many sustainability performance measurements, we have not seen many performance measurements developed for the food supply chain specifically. According to FAO (2012, p. 9), '[n]either a commonly accepted set of indicators that have to be taken into account when measuring sustainability performance, nor widely accepted definitions of the minimum requirements that would allow a company to qualify as "sustainable", exist'. We have seen more works focusing on environmental and economic aspects, but not on the social aspect (Martins *et al.*, 2019). Eskandarpour *et al.* (2015) argued that the lack of social aspect is more difficult to quantify social factors. In response to that gap, Bloemhof *et al.* (2015) started an international project named Step Change in Agri-food Logistics Ecosystems (SCALE), which aims to develop tools and frameworks for sustainable food supply chains. The project identified internal drivers (e.g. cost, efficiency and brand reputation), external drivers (e.g. policy and regulation), and barriers (e.g. the habit of short-term contracts) that are important for sustainable food supply chains. However, the project was based on the context of the Netherlands, France and the UK. Thus, it might not be relevant to some underdeveloped or developing countries. Another limitation was that the project did not provide a method for calculating the overall sustainability score, which supports communication among stakeholders.

6.6. Trends in Sustainable Food Supply Chains

This section discusses trends that affect the development of a sustainable food supply chain. Particularly, it discusses sustainable food supply chain modelling, digitization, robotics and autonomous systems, and training and education.

6.6.1. Sustainable food supply chain modelling

As the food industry has grown exponentially to satisfy the increasing demand, researchers and policymakers have sought models that make food supply chains more sustainable. One of the most difficult is finding a solution that compromises short-term objectives (e.g. consumer demand, low costs) and long-term objectives (e.g. zero-carbon emission). At the current state, natural resources for food production are not sustainable, but affordable food products are an urgent requirement. Literature provides several optimization models to aid planning activities along the food supply chain. Ahumada and Villalobos (2009) conducted a comprehensive review of planning models in the food supply chain. The authors categorized selected papers into three groups: perishable and non-perishable products; strategic, tactical or operational decision-making levels; and production, harvest, storage and distribution functions. Ahumada and Villalobos (2009) stated that most of the literature focuses on one planning aspect and lacks integrated planning models in the food supply chain. It is because integrated models are complex and lack real-world datasets. Another finding is that although perishable products have a stochastic shelf-life, planning models normally fail to incorporate the stochastic characteristic. Moreover, not many papers focus on the operational planning level.

Zhu *et al.* (2018) reviewed 83 papers that used mathematical modelling to address sustainable food supply chain issues. They called for consideration of globalization, consumer preferences, regional food hubs, temperature-controlled storage, transportation and distribution. Specifically, social aspects, such as farmers' welfare, animal welfare and food insecurity are crucial factors for further exploration. Kamble *et al.* (2020) highlighted the importance of a data-driven sustainable food supply chain. They reviewed 84 academic papers to understand the level of analytics used (descriptive, predictive and prescriptive), sustainable objectives attained (social, environmental and economic), supply chain processes, and supply chain resources. They suggested that researchers should focus on improving supply chain visibility, exploring the possibilities of implementing the Internet of Things, using blockchain technologies, and developing big-data capabilities. These activities can make a food supply chain more sustainable.

6.6.2. Digitization

To ensure consumer safety and food quality, regulators have enforced a wide range of standards and certifications critical in managing safety, quality and transparency in the food industry. However, responsibility for food quality belongs to all companies within the food supply chain, not just specific companies. Many researchers have advocated integrating digitization in the food supply chain (Duong *et al.*, 2020b). Such integration benefits traceability (Garcia-Torres *et al.*, 2019) and legal culpability (Dubey *et al.*, 2020).

The availability and advantages of digital technologies and increasing big data analysis capabilities have enabled food supply chains to be more efficient, cost-effective and responsive to market demands (Kittipanya-ngam and Tan, 2020). For example, traceability, transparency and food safety can be improved with digital platforms (Bouzembrak *et al.*, 2019; Creydt and Fischer, 2019). Extensive literature has discussed the benefits of digitization in the food industry. Zhao *et al.* (2019) reviewed advances and main applications of blockchain technology in the food supply chain. The authors highlighted four main benefits of blockchain including traceability, information security, sustainable water management, and manufacturing. Belaud *et al.* (2019) designed an approach integrating big data to improve sustainable management for agricultural waste. These recent examples represent huge benefits of digitization for sustainable food supply chains.

However, it is noted that while digital technologies have been available for a long time, the adoption of digital technologies in the food supply chain has just been highlighted recently. There are two main reasons for this delay (Annosi *et al.*, 2020): there is a mismatch between the adopters (e.g. farmers) and digitization providers. To successfully adopt digital technologies, the adopters have to invest heavily in labour, skills and technologies; on the other hand, the providers seem to ignore the complexity of business models in the food industry and the technological readiness of adopters. Secondly, the complex and long food supply chain causes significant differences among supply chain partners in terms of knowledge, experience, attitudes, education and awareness, preventing the adoption of digital technologies in the food supply chain.

6.6.3. Robotics and autonomous systems

Researchers also call for adopting robotics and autonomous systems (RAS) in the food industry (Rehman *et al.*, 2019). RAS is not a new concept, and it has been used widely in the food industry since the end of the 20th century (Miranda *et al.*, 2019). Harvesting robots for the greenhouse are projected to be used at least in 20% of future harvesting operations (Suprem *et al.*, 2013). In food processing factories, collaborative robots, which work side-by-side with operators, can increase safety and productivity, especially for hazardous human tasks (Guiochet *et al.*, 2017). RAS also is used regularly for milking animals, and it is forecast that RAS could be adopted in half of the operations in the dairy industry (Beekman and Bodde, 2015).

From the sustainability perspective, RAS can benefit the food industry through five major aspects: food quality, food safety, food waste, supply chain efficiency, and supply chain analysis. First, image-processing methods or computer-vision technologies are used to examine food quality at farms, factories or delivery truck (Haass *et al.*, 2015; Trivelli *et al.*, 2019). Second, RFID and wireless sensor networks can be utilized to collect humidity and temperature during food transportation or food storage (Alfian *et al.*, 2017). Third, detection technologies or sensors can help monitor and minimize food waste along the supply chain. Fourth, RAS collects and shares information, which is critical in measuring sustainability indicators (Bouzembrak *et al.*, 2019). Finally, results from such measurement could be used for analysis and benchmarking, and, in return, support the implementation of sustainability strategies. However, the adoption of RAS in the food industry also encounters many challenges, such as data availability, cybersecurity, skill capability, and financial cost. We invite readers to visit Duong *et al.*'s (2020b) 'A review of robotics and autonomous systems in the food industry: from the supply chains perspective' for a comprehensive review of the adoption of RAS in the food supply chain.

6.6.4. Training and education

Achieving sustainability is a great challenge for humanity in general and the food industry in particular. A key characteristic of the food industry is that it involves many people. Thus, it is vital to ensure that employees in the food industry have skills and awareness by providing necessary training and education. A recent survey showed that food waste negatively impacts nine out of ten employees of food-to-go businesses daily at their business (Too Good To Go, 2020). However, food firms have not taken this issue seriously. Only 45% of employees surveyed said they receive sustainability awareness and training is very limited (Too Good To Go, 2020).

In this respect, training and education are essential to prepare employees for a path to sustainability. The food industry is extensive; thus, training and education should reach all involved people with different backgrounds and cultures. The contribution of governments and employers has a direct relationship to the success of training and education programmes.

Medeiros (2017) summarized specific themes that should be included in training and education for a sustainable food supply chain. These themes include conservation and environmental protection, sustainable consumption and production, environmentally sustainable technologies, sustainable cities, human rights, healthcare, gender equality, and poverty reduction. Upon on the background and culture, firms should carefully plan their training and education programmes to enhance their effectiveness.

6.7. Conclusion

The success of a sustainable food supply chain is uncertain unless more efforts are made to ensure all involved partners understand issues and challenges. More effort is needed to redesign the food supply chain. Sustainable food supply chains provide enough food for the increasing demand and make sure that future generations still have the resources they need. This requires a trade-off among three dimensions of sustainability: environment, economy and society. This chapter reviewed a list of issues and challenges for sustainable supply chains. It then discussed future research trends, including sustainable food supply chain modelling, digitization, robotics and autonomous systems, and training and education.

References

Ahumada, O. and Villalobos, J.R. (2009) Application of planning models in the agri-food supply chain: a review. *European Journal of Operational Research* 196, 1–20.

Alfian, G., Rhee, J., Ahn, H., Lee, J., Farooq, U., Ijaz, M.F. and Syaekhoni, M.A. (2017) Integration of RFID, wireless sensor networks, and data mining in an e-pedigree food traceability system. *Journal of Food Engineering* 212, 65–75.

Annosi, M.C., Brunetta, F., Capo, F. and Heideveld, L. (2020) Digitalization in the agri-food industry: the relationship between technology and sustainable development. *Management Decision* 58, 1737–1757.

Assefa, T.T., Meuwissen, M.P.M. and Lansink, A.G.J.M.O. (2015) Price volatility transmission in food supply chains: a literature review. *Agribusiness* 31, 3–13.

Beekman, J. and Bodde, R. (2015) Milking automation is gaining popularity. Dairy Global. Available at: https://www.dairyglobal.net/Milking/Articles/2015/1/Milking-automation-is-gaining-popularity-1568767W/ (accessed 22 June 2020).

Belaud, J.-P., Prioux, N., Vialle, C. and Sablayrolles, C. (2019) Big data for agri-food 4.0: application to sustainability management for by-products supply chain. *Computers in Industry* 111, 41–50.

Béné, C. (2020) Resilience of local food systems and links to food security – a review of some important concepts in the context of COVID-19 and other shocks. *Food Security* 12, 805–822.

Bloemhof, J.M., Vorst, J.G.A.J., van der, Bastl, M. and Allaoui, H. (2015) Sustainability assessment of food chain logistics. *International Journal of Logistics Research and Applications* 18, 101–117.

Bouzembrak, Y., Klüche, M., Gavai, A. and Marvin, H.J.P. (2019) Internet of Things in food safety: literature review and a bibliometric analysis. *Trends in Food Science & Technology* 94, 54–64.

Bush, C. (2010) Sustainable sourcing: a new approach to high performance in supply chain management. Accenture. Available at: https://www.criticaleye.com/insights-servfile.cfm?id=723 (accessed 3 August 2021).

Bustos, C.A. and Moors, E.H.M. (2018) Reducing post-harvest food losses through innovative collaboration: insights from the Colombian and Mexican avocado supply chains. *Journal of Cleaner Production* 199, 1020–1034.

Chardine-Baumann, E. and Botta-Genoulaz, V. (2014) A framework for sustainable performance assessment of supply chain management practices. *Computers & Industrial Engineering* 76, 138–147.

Cole, R., Stevenson, M. and Aitken, J. (2019) Blockchain technology: implications for operations and supply chain management. *Supply Chain Management: An International Journal* 24, 469–483.

Creydt, M. and Fischer, M. (2019) Blockchain and more: algorithm driven food traceability. *Food Control* 105, 45–51.

Dora, M., Wesana, J., Gellynck, X., Seth, N., Dey, B. and De Steur, H. (2020) Importance of sustainable operations in food loss: evidence from the Belgian food processing industry. *Annals of Operations Research* 290, 47–72.

Dubey, R., Gunasekaran, A., Childe, S.J., Papadopoulos, T., Luo, Z. and Roubaud, D. (2020) Upstream supply chain visibility and complexity effect on focal company's sustainable performance: Indian manufacturers' perspective. *Annals of Operations Research* 290, 343–367.

Duong, L.N.K., Wood, L.C. and Wang, W.Y.C. (2018) Effects of consumer demand, product lifetime, and substitution ratio on perishable inventory management. *Sustainability* 10(5), 1–17.

Duong, L.N.K., Wood, L.C. and Wang, W.Y.C. (2020a) Inventory management of perishable health products: a decision framework with non-financial measures. *Industrial Management & Data Systems* 120, 987–1002.

Duong, L.N.K., Al-Fadhli, M., Jagtap, S., Bader, F., Martindale, W., Swainson, M. and Paoli, A. (2020b) A review of robotics and autonomous systems in the food industry: from the supply chains perspective. *Trends in Food Science & Technology* 106, 355–364.

Eskandarpour, M., Dejax, P., Miemczyk, J. and Péton, O. (2015) Sustainable supply chain network design: an optimization-oriented review. *Omega* 54, 11–32.

FAO (2012) *FAO Statistical Yearbook 2012: World Food & Agriculture*. Food and Agriculture Organization of the United Nations. Available at: http://www.fao.org/3/i2490e/i2490e00. htm (accessed 6 January 2021).

FAO (2017) Food loss and food waste. *Food and Agriculture Organization of the United Nations*. Available at: http://www.fao.org/food-loss-and-food-waste/en/ (accessed 4 January 2021).

Garcia-Torres, S., Albareda, L., Rey-Garcia, M. and Seuring, S. (2019) Traceability for sustainability – literature review and conceptual framework. *Supply Chain Management: An International Journal* 24, 85–106.

Gimenez, C., Sierra, V. and Rodon, J. (2012) Sustainable operations: their impact on the triple bottom line. *International Journal of Production Economics* 140, 149–159.

Guiochet, J., Machin, M. and Waeselynck, H. (2017) Safety-critical advanced robots: a survey. *Robotics and Autonomous Systems* 94, 43–52.

Haass, R., Dittmer, P., Veigt, M. and Lütjen, M. (2015) Reducing food losses and carbon emission by using autonomous control – a simulation study of the intelligent container. *International Journal of Production Economics* 164, 400–408.

Jabbour, C.J.C., de Sousa Jabbour, A.B.L. and Sarkis, J. (2019) Unlocking effective multi-tier supply chain management for sustainability through quantitative modeling: lessons learned and discoveries to be made. *International Journal of Production Economics* 217, 11–30.

Jones, T. (2020) Future food standards – it's not just about chlorinated chicken. *The Grocer*. Available at: https://www.thegrocer.co.uk/sourcing/future-food-standards-its-not-just-about-chlorinated-chicken/649735.article (accessed 3 January 2021).

Kamble, S.S., Gunasekaran, A. and Gawankar, S.A. (2020) Achieving sustainable performance in a data-driven agriculture supply chain: a review for research and applications. *International Journal of Production Economics* 219, 179–194.

Khanal, U., Wilson, C., Hoang, V.-N. and Lee, B.L. (2019) Autonomous adaptations to climate change and rice productivity: a case study of the Tanahun district, Nepal. *Climate and Development* 11, 555–563.

Kittipanya-ngam, P. and Tan, K.H. (2020) A framework for food supply chain digitalization: lessons from Thailand. *Production Planning & Control* 31, 158–172.

Kleindorfer, P.R., Singhal, K. and Wassenhove, L.N.V. (2005) Sustainable operations management. *Production and Operations Management* 14, 482–492.

Kong, D., Shi, L. and Yang, Z. (2019) Product recalls, corporate social responsibility, and firm value: evidence from the Chinese food industry. *Food Policy* 83, 60–69.

Lee, H.L., Padmanabhan, V. and Whang, S. (1997) Information distortion in a supply chain: the bullwhip effect. *Management Science* 43, 546–558.

Liu, R., Gao, Z., Nayga, R.M., Snell, H.A. and Ma, H. (2019) Consumers' valuation for food traceability in China: Does trust matter? *Food Policy* 88, 101768.

Maier, P., Klein, O. and Schumacher, K.P. (2020) Ecological benefits through alternative food networks? Prospects of regional barley-malt-beer value chains in Bavaria, Germany. *Journal of Cleaner Production* 265, 121848.

Mangla, S.K., Sharma, Y.K., Patil, P.P., Yadav, G. and Xu, J. (2019) Logistics and distribution challenges to managing operations for corporate sustainability: study on leading Indian dairy organizations. *Journal of Cleaner Production* 238, 117620.

Martins, C.L., Melo, M.T. and Pato, M.V. (2019) Redesigning a food bank supply chain network in a triple bottom line context. *International Journal of Production Economics* 214, 234–247.

Medeiros, C.O. (2017) Sustainability challenges and educating people involved in the agrofood sector. In: Bhat, R. (ed.) *Sustainability Challenges in the Agrofood Sector*. John Wiley & Sons, Oxford, UK, 660–674.

Mentzer, J.T., DeWitt, W., Keebler, J.S., Min, S., Nix, N.W., Smith, C.D. and Zacharia, Z.G. (2001) Defining supply chain management. *Journal of Business Logistics* 22, 1–25.

Miranda, J., Pérez-Rodríguez, R., Borja, V., Wright, P.K. and Molina, A. (2019) Sensing, smart and sustainable product development (S3 product) reference framework. *International Journal of Production Research* 57, 4391–4412.

Nardi, V.A.M., Teixeira, R., Ladeira, W.J. and de Oliveira Santini, F. (2020) A meta-analytic review of food safety risk perception. *Food Control* 112, 107089.

Patra, D., Leisnham, P.T., Tanui, C.K. and Pradhan, A.K. (2020) Evaluation of global research trends in the area of food waste due to date labeling using a scientometrics approach. *Food Control* 115, 107307.

Pohlmann, C.R., Scavarda, A.J., Alves, M.B. and Korzenowski, A.L. (2020) The role of the focal company in sustainable development goals: a Brazilian food poultry supply chain case study. *Journal of Cleaner Production* 245, 118798.

Rehman, T.U., Mahmud, Md.S., Chang, Y.K., Jin, J. and Shin, J. (2019) Current and future applications of statistical machine learning algorithms for agricultural machine vision systems. *Computers and Electronics in Agriculture* 156, 585–605.

Reynolds, J. and Dolasinski, M.J. (2019) Systematic review of industry food safety training topics & modalities. *Food Control* 105, 1–7.

Schanes, K., Dobernig, K. and Gözet, B. (2018) Food waste matters: a systematic review of household food waste practices and their policy implications. *Journal of Cleaner Production* 182, 978–991.

Seuring, S. and Müller, M. (2008) From a literature review to a conceptual framework for sustainable supply chain management. *Journal of Cleaner Production* 16, 1699–1710.

Sibhatu, K.T. and Qaim, M. (2018) Review: meta-analysis of the association between production diversity, diets, and nutrition in smallholder farm households. *Food Policy* 77, 1–18.

Suprem, A., Mahalik, N. and Kim, K. (2013) A review on application of technology systems, standards and interfaces for agriculture and food sector. *Computer Standards & Interfaces* 35, 355–364.

Taghikhah, F., Voinov, A., Shukla, N., Filatova, T. and Anufriev, M. (2021) Integrated modeling of extended agro-food supply chains: a systems approach. *European Journal of Operational Research* 288, 852–868.

Terazono, E. (2020) Food price rally sparks warnings of pressure on developing countries. *Financial Times*. Available at: https://www.ft.com/content/20a5e763-4d2f-4872-a1f9-7852a5be6d68 (accessed 14 January 2021).

Terazono, E. and Evans, J. (2020) How coronavirus is affecting pasta's complex supply chain. *Financial Times*. Available at: https://www.ft.com/content/5456bc24-6dd4-11ea-9bca-bf503995cd6f (accessed 27 December 2020).

Too Good To Go (2020) Food waste has got to go. Available at: https://toogoodtogo.co.uk/en-gb/lp/b2b/foodtogo-industry-report (accessed 13 January 2021).

Trivelli, L., Apicella, A., Chiarello, F., Rana, R., Fantoni, G. and Tarabella, A. (2019) From precision agriculture to Industry 4.0: unveiling technological connections in the agrifood sector. *British Food Journal* 121, 1730–1743.

United Nations (2005) World summit outcome: resolution adopted by the general assembly. Available at: https://www.un.org/en/development/desa/population/migration/generalassembly/docs/globalcompact/A_RES_60_1.pdf (accessed 28 December 2020).

van der Vorst, J.G.A.J., Tromp, S.O. and van der Zee, D.-J. (2009) Simulation modelling for food supply chain redesign: integrated decision making on product quality, sustainability and logistics. *International Journal of Production Research* 47, 6611–6631.

Wallach, O. (2020) From bean to brew: the coffee supply chain. *Visual Capitalist*. Available at: https://www.visualcapitalist.com/from-bean-to-brew-the-coffee-supply-chain/ (accessed 02 January 2021).

WCED (1987) *Our Common Future*. Oxford University Press, Oxford, UK.

Wossen, T., Berger, T., Haile, M.G. and Troost, C. (2018) Impacts of climate variability and food price volatility on household income and food security of farm households in East and West Africa. *Agricultural Systems* 163, 7–15.

Zhao, G., Liu, S., Lopez, C., Lu, H., Elgueta, S., Chen, H. and Boshkoska, B.M. (2019) Blockchain technology in agri-food value chain management: a synthesis of applications, challenges and future research directions. *Computers in Industry* 109, 83–99.

Zhu, Z., Chu, F., Dolgui, A., Chu, C., Zhou, W. and Piramuthu, S. (2018) Recent advances and opportunities in sustainable food supply chain: a model-oriented review. *International Journal of Production Research* 56, 5700–5722.

7 Transformational Systems and Resilience in Food Manufacturing

LINH DUONG

7.1. Introduction

At the end of December 2020, just a few days before Christmas, France closed its border to all drivers from the UK without notice as a measure to control the spread of a new COVID-19 variant (*Financial Times*, 2020). Soon after this, there was a long queue of trucks backing up outside the Channel port. This drastic action caused painful losses for UK businesses. Perishable products like fish and seafood had to be destroyed before they arrived with consumers. At the same time, non-perishable goods industries faced the risk of not holding enough stock ahead of the new year. Additionally, truck drivers were exhausted as they arrived home for Christmas.

This reminds us that we live in an uncertain world with many disruptive events caused by nature and humans. Once a disruption happens, it causes inconvenience for businesses as they might not deliver or receive the products or service as planned (Ivanov, 2020). As supply chains become more complex and longer, it is not surprising that the exposure to disruption has increased (Scheibe and Blackhurst, 2018). Specifically, as the food supply chains have expanded globally, the risk for the food supply chain increases significantly.

This chapter aims to introduce the complex topic of resilience, especially for the food supply chain. It defines key concepts relating to disruption and resilience. In particular, it discusses resilience issues relating to food supply chains. Then it discusses research trends in this field.

7.2. Supply Chain Disruption

Supply chain disruptions are defined as 'unplanned and unanticipated events that disrupt the normal flow of goods and materials within a supply chain' (Craighead *et al.*, 2007, p. 132). Disruption can happen due to many reasons such as delays

© CAB International 2022. *Food Industry 4.0: Unlocking Advancement Opportunities in the Food Manufacturing Sector* (W. Martindale *et al.*)
DOI: 10.1079/9781789248593.0007

Box 7.1. Impact of COVID-19 in the Food Supply Chain (Barman *et al.*, 2021).

Impact of COVID-19 in the Food Supply Chain

The COVID-19 pandemic is a huge disruption for the food supply chain. It affects both supply and demand sides, from the farm to the consumer. Consequently, the security of food is also at risk. Extra requirements on sanitization, social distancing, cleaning and disinfecting the facilities stress food production, food distribution, and disconnect the fresh food market.

One of the priorities of food companies is protecting the health of labour and having enough labour for the operations. Due to disruption in transportation, raw materials could not be transferred to food factories, e.g. many farmers had to dump milk. Maintenance of the movement of product is an essential factor in the food industry as the biggest issue in the food supply chain is having enough raw materials and ensuring the distribution of final products to consumers.

The COVID-19 pandemic increases the demand for food products. The demand for bread and vegetables increased 76% and 52%, respectively, in the week when the pandemic was declared. Higher demand leads to empty racks and a higher cost of food items. Consequently, many governments struggled under financial pressure due to the economic loss and rising costs for public help programmes.

Issues in both supply and demand lead to food insecurity as many people have more difficulties fulfilling their fundamental food requirements. There is an increasing number of countries having a growing level of food insecurity. In the UK, the government extended the free school meals scheme to help pupils that miss out on a healthy school meal.

---

in processing, disasters, systems breakdown, inaccurate forecast, inventory, or capacity (Chopra and Sodhi, 2014). Once it happens, it may prevent a company or a whole supply chain from operating at a planned state. As mentioned in previous chapters, the food supply chain has become complex and global; the cost of any disruption event can be expensive. In addition, supply chain disruptions have significant impacts on inventory performance (Sarkar and Kumar, 2015), firms' manufacturing (Fartaj *et al.*, 2020) and firms' profit (Gaur *et al.*, 2020). For example, the COVID-19 pandemic caused a shortage of labour and logistics disruption as governments impose a lockdown to control the transmission of COVID-19 (Duong *et al.*, 2020). Consequently, there are many temporary shortages of some food product ranges in supermarkets. Researchers have proposed models and managerial processes to identify, assess and mitigate supply chain disruptions (Azadegan *et al.*, 2020; Duong and Chong, 2020).

7.3. Supply Chain Risk

On the other side, supply chain risk is often considered as the cause of supply chain disruption (DuHadway *et al.*, 2019). Supply chain risk has been defined in many ways. It is considered as 'any risks for the information, material and product flows from original suppliers to the delivery of the final product for the end-user' (Jüttner *et al.*, 2003, p. 200); or 'the potential variation of outcomes that influence the decrease of value added at any activity cell in a chain'

(Bogataj and Bogataj, 2007, p. 291); or 'an individual's perception of the total potential loss associated with the disruption of supply of a particular purchased item from a particular supplier' (Ellis *et al.*, 2010, p. 36). Taking a comprehensive review, Ho *et al.* (2015, p. 5035) defined it as 'the likelihood and impact of unexpected macro- and/or micro-level events or conditions that adversely influence any part of a supply chain leading to operational, tactical, or strategic level failures or irregularities'. The discussion in this chapter is based on this definition as it reflects the comprehensive view of supply chain risk. Supply chain risks, when they happen, have impacts on the performance of the whole supply chain (Ritchie and Brindley, 2007; Munir *et al.*, 2020). Researchers have considered supply chain risk management as crucial strategic and tactical activities (Garvey and Carnovale, 2020). Consequently, many studies proposed models and frameworks for supply chain risk management. Some examples include risk assessment (Pettit *et al.*, 2010), risk classification (Ho *et al.*, 2015), collaboration mechanisms (Duong and Chong, 2020), as well as decision modelling (Behzadi *et al.*, 2018).

While supply chain risk management has received much attention, Garvey *et al.* (2015) noted that most researchers focused on the firm's horizon. However, there are growing calls for understanding the impacts of external factors on supply chain risk management (Carter *et al.*, 2015). For example, Dixit *et al.* (2020) measured the risk of a supply chain based on its network structural parameters. Yan *et al.* (2020) designed a co-ordination contract for a decentralized fresh agricultural product supply chain. The idea of these is that considering the external factors of a supply chain can improve the profit for the whole chain.

Although supply chain risk management is a common research topic, there is little attention in the food industry (Pereira *et al.*, 2020). Christopher *et al.* (2011) draw attention to developing risk assessment methods and mitigation strategies from the whole supply chain perspective. The majority of the literature focuses on the food industry in a specific country. Nyamah *et al.* (2017) examined the probability and consequences of risk factors that affect the food supply chains in Ghana. Leat and Revoredo-Giha (2013) identified the main risks and challenges in Scotland's pork supply chains. While these studies provide insightful managerial implications, the results might be area-biased.

7.4. Supply Chain Resilience

Once a disruption happens, a firm's performance might be affected. For example, it has difficulties delivering products to consumers because there are not enough raw materials for production or the distribution network is interrupted. Resilience can be defined as a firm's capabilities to respond to and recover from a disruption to the original state of performance, or even to a more desirable performance level (Christopher and Peck, 2004). It is presented by 'four Rs': resourcefulness, robustness, recovery and review (Pettit *et al.*, 2013). According to Ponomarov and Holcomb (2009), supply chain resilience helps firms prepare for, respond to and recover from a disruption.

In the last decades, we have witnessed an increasing number of disruptions leading to strong attention on supply chain resilience. Inevitably, a wide range of strategies has been suggested in the literature for building a resilient supply chain (Urciuoli *et al.*, 2014). In this aspect, collaboration, agility, flexibility and redundancy strategies are highlighted as the most important capabilities to improve supply chain resilience (Jain *et al.*, 2017; Simchi-Levi *et al.*, 2018). The majority of literature focuses on strategies for building resilience capabilities at a particular firm. However, the effectiveness of such strategies requires alignment or collaboration across supply chain partners (Ali *et al.*, 2017).

7.5. How to Build a Resilient Food Supply Chain

Another approach to achieving supply chain resilience is to assume that resilience has two critical components: resistance and recovery (Christopher, 2013). Resistance is defined as the robustness to avoid turbulence; recovery refers to the ability to bounce back quickly after the disruption. From this perspective, Christopher (2013) proposed four key elements of supply chain resilience. We will discuss each of them in turn with a focus on the agri-food supply chain.

7.5.1. Supply chain (re-)engineering

Many risk sources come from the design of the supply chain. The cost factor drives many decisions in supply chain management. Thus we have seen an increasing number of big and centralized warehouses reduce operational costs. There is also a trend in finding sources from developing countries like China, India or Vietnam, where the labour cost and the overall costs are lower. However, this practice could increase supply chain risk. If the supply chain could develop multiple suppliers, it can avoid reliance on a single source. With multi-sourcing, a food company can avoid the risk of not having enough materials as well as the issues of quality. For example, the melamine contamination in the Chinese dairy industry led to more than 300,000 victims in 2008 (Marucheck *et al.*, 2011). In this context, Bottani *et al.* (2019) developed a

---

Box 7.2. Global food prices (Terazono, 2021).

Global Food Prices

Global food prices have increased by the biggest margin in many years. The easing of COVID-19 restrictions and the rise in transport costs push food prices higher. In addition, bad weather in Brazil, an exporter of corn and soybean, and rising demand are expected to push prices even higher.

This high increase hits poor countries reliant on imports of food products. In West Africa, the price is up 40% over a five-year average. Vulnerable countries such as Lebanon and Syria are struggling with food prices with rises of up to over 200%.

model that maximizes the total profit and minimizes the total supply chain lead time. The model is used to design a resilient food supply chain that uses a multiple sourcing policy for multiple products. Chopra and Sodhi (2014) suggested that the higher number of facilities leads to a higher cost of resources and lower disruption costs. Thus, the supply chain design might need a trade-off decision between the number of facilities and resilience level.

7.5.2. Supply chain collaboration

Because there are many partners involved in the food supply chain, the collaboration between them is another main element of resilience. According to Cao *et al.*, supply chain collaboration is 'a long-term partnership process where supply chain partners with common goals work closely together to achieve mutual advantages that are greater than the firms would achieve individually' (2010, p. 6616). Supply chain collaboration happens in many forms and can be grouped into seven categories: joint practices, contractual and economics practices, technological and information sharing practices, assessment practices, relationship management, governance practices, and supply chain design (Cloutier *et al.*, 2020). Supply chain collaboration can deliver substantial advantages such as risk sharing, information sharing, and service improvements (Blome *et al.*, 2014; Chen *et al.*, 2017; Allaoui *et al.*, 2019). Duong and Chong (2020) identified that information sharing and technology, trust, culture, stakeholders, divergent goals, flexibility, knowledge and experience, market factor, measurement issues, resources, and visibility influence collaboration. However, information sharing and technology, knowledge and experience have received less attention (Cao *et al.*, 2010). Collaboration is not merely transactions; it is more about communication, sharing information, and sharing knowledge for sustainable competitive advantage (Kao and Wu, 2016).

7.5.3. Supply chain risk management culture

A failure in supply chain management can lead to massive damage; it is vital that business leaders recognize the importance of risk management and provide leadership in risk management. Supply chain risk management culture represents the behaviour and belief of staff and managers in mitigating and dealing with disruption (Kumar and Anbanandam, 2020). Business managers should review regular reports on risk profile in making a decision (Christopher and Peck, 2004), which should not negatively affect the firm's abilities to deal with disruption or make the firm more vulnerable. Such a culture requires leadership and a continuity team to engage with all supply chain partners and monitor and control the process.

Building a continuity team is an important task given the growing concern of food safety and food quality. Zhang *et al.* (2011) developed a diagnosis model to provide pre-warning signals of quality in the food industry.

> **Box 7.3.** Business continuity. (From BSI, n.d.)
>
> **Business Continuity**
>
> Business continuity is a growing concern and rising discipline in the food industry. It is still a new concept with many companies still lacking proactive and preventative crisis management systems. The limited resource and expertise slow business continuity management implementation. One important approach to achieve business continuity is the use of international standards. This has been driven by several Global Food Safety Initiative standards such as Safe Quality Food Initiative and British Retail Consortium. They determine requirements for setting up business continuity plan and guide companies to recovery and restoration of major functions.

The creation of a continuity team can positively affect a firm's attitudes and encourage actions towards risks. This also enhances the awareness of supply chain risks that hinder the flow of products.

7.5.4. Agility

To reduce the risk of a supply chain, its agility is vital, as the more agile it is, the quicker it recovers from disruption (Christopher, 2013). Agility is defined as 'the ability to cope with unexpected challenges, to survive unprecedented threats of business environment, and to take advantage of changes as opportunities' (Sharifi and Zhang, 1999, p. 9). There is rich literature discussing supply chain agility. Charles *et al.* (2010) measured agility in five dimensions: visibility, velocity, flexibility, reliability and effectiveness. Gligor and Holcomb (2012a) described supply chain agility in terms of flexibility, responsiveness, customization, change as opportunity, integration, organizational structure, mobilization of core competencies, and speed. Eckstein *et al.* (2015, p. 3029) comprehensively described agility as:

> the ability of the firm to sense short-term, temporary changes in supply chain and market environment (e.g. demand fluctuations, supply fluctuations, changes in suppliers' delivery times), and to rapidly and flexibly respond to those changes within the existing supply chain (e.g. reducing replacement times of materials, reducing manufacturing throughput times, adjusting delivery capacities).

Due to its vital role, many studies have investigated antecedents and enablers of supply chain agility. Blome *et al.* (2013) suggested that supply- and demand-side competence can be transformed via supply chain agility into superior performance. Gligor and Holcomb (2012b) indicate that co-operation, co-ordination and communication directly impact supply chain agility. Gligor *et al.* (2016) revealed that environment, market orientation and supply chain are antecedents of agility. Recently, Dubey *et al.* (2018) utilized the resource-based view theory and suggested that information sharing and supply chain connectivity resources influence supply chain agility.

7.6. Conclusion

The increasing number of disruptive events and changes in consumer behaviour have exposed supply chains to more vulnerabilities and damages. It is vital for companies, especially in the food industry, to build resilient supply chains to resist and recover quickly from any disruptive event. Success in building a resilient supply chain benefits economic objectives and contributes to society. During the COVID-19 pandemic, Sainsbury's, Asda and other supermarkets have quickly promoted click-and-collect service. This brings convenience for customers who want to keep social distancing and secures profitability for the supermarkets. Technologies have a more and more important role in building resilient supply chains. Future research can investigate the role of technologies and how to use technologies in building resilient supply chains.

References

Ali, I., Nagalingam, S. and Gurd, B. (2017) Building resilience in SMEs of perishable product supply chains: enablers, barriers and risks. *Production Planning & Control* 28, 1236–1250.

Allaoui, H., Guo, Y. and Sarkis, J. (2019) Decision support for collaboration planning in sustainable supply chains. *Journal of Cleaner Production* 229, 761–774.

Azadegan, A., Mellat Parast, M., Lucianetti, L., Nishant, R. and Blackhurst, J. (2020) Supply chain disruptions and business continuity: an empirical assessment. *Decision Sciences* 51, 38–73.

Barman, A., Das, R. and De, P.K. (2021) Impact of COVID-19 in the food supply chain: disruptions and recovery strategy. *Current Research in Behavioral Sciences* 2, 100017.

Behzadi, G., O'Sullivan, M.J., Olsen, T.L. and Zhang, A. (2018) Agribusiness supply chain risk management: a review of quantitative decision models. *Omega* 79, 21–42.

Blome, C., Schoenherr, T. and Rexhausen, D. (2013) Antecedents and enablers of supply chain agility and its effect on performance: a dynamic capabilities perspective. *International Journal of Production Research* 51, 1295–1318.

Blome, C., Paulraj, A. and Schuetz, K. (2014) Supply chain collaboration and sustainability: a profile deviation analysis. *International Journal of Operations & Production Management* 34, 639–663.

Bogataj, D. and Bogataj, M. (2007) Measuring the supply chain risk and vulnerability in frequency space. *International Journal of Production Economics* 108, 291–301.

Bottani, E., Murino, T., Schiavo, M. and Akkerman, R. (2019) Resilient food supply chain design: modelling framework and metaheuristic solution approach. *Computers & Industrial Engineering* 135, 177–198.

BSI (n.d.) Fail to prepare, prepare to fail – business continuity management in the food industry [online]. Available at: https://www.bsigroup.com/LocalFiles/en-GB/food-and-drink/documents/food-business-continuity-management.pdf (accessed 7 June 2021).

Cao, M., Vonderembse, M.A., Zhang, Q. and Ragu-Nathan, T.S. (2010) Supply chain collaboration: conceptualisation and instrument development. *International Journal of Production Research* 48, 6613–6635.

Carter, C.R., Rogers, D.S. and Choi, T.Y. (2015) Toward the theory of the supply chain. *Journal of Supply Chain Management* 51, 1–23.

Charles, A., Lauras, M. and van Wassenhove, L.N. (2010) A model to define and assess the agility of supply chains: building on humanitarian experience. *International Journal of Physical Distribution & Logistics Management* 40 (8/9), 722–741.

Chen, L., Zhao, X., Tang, O., Price, L., Zhang, S. and Zhu, W. (2017) Supply chain collaboration for sustainability: a literature review and future research agenda. *International Journal of Production Economics* 194, 73–87.

Chopra, S. and Sodhi, M.S. (2014) Reducing the risk of supply chain disruptions. *MIT Sloan Management Review* 55, 72–80.

Christopher, M. (2013) *Logistics and Supply Chain Management,* 4th edn. Pearson, London.

Christopher, M. and Peck, H. (2004) Building the resilient supply chain. *International Journal of Logistics Management* 15, 1–14.

Christopher, M., Mena, C., Khan, O. and Yurt, O. (2011) Approaches to managing global sourcing risk. *Supply Chain Management: An International Journal* 16, 67–81.

Cloutier, C., Oktaei, P. and Lehoux, N. (2020) Collaborative mechanisms for sustainability-oriented supply chain initiatives: state of the art, role assessment and research opportunities. *International Journal of Production Research* 58, 5836–5850.

Craighead, C.W., Blackhurst, J., Rungtusanatham, M.J. and Handfield, R.B. (2007) The severity of supply chain disruptions: design characteristics and mitigation capabilities. *Decision Sciences* 38, 131–156.

Dixit, V., Verma, P. and Tiwari, M.K. (2020) Assessment of pre and post-disaster supply chain resilience based on network structural parameters with CVaR as a risk measure. *International Journal of Production Economics* 227, 107655.

Dubey, R., Altay, N., Gunasekaran, A., Blome, C., Papadopoulos, T. and Childe, S.J. (2018) Supply chain agility, adaptability and alignment: empirical evidence from the Indian auto components industry. *International Journal of Operations & Production Management* 38, 129–148.

DuHadway, S., Carnovale, S. and Hazen, B. (2019) Understanding risk management for intentional supply chain disruptions: risk detection, risk mitigation, and risk recovery. *Annals of Operations Research* 283, 179–198.

Duong, L.N.K. and Chong, J. (2020) Supply chain collaboration in the presence of disruptions: a literature review. *International Journal of Production Research* 58, 3488–3507.

Duong, L.N.K., Al-Fadhli, M., Jagtap, S., Bader, F., Martindale, W., Swainson, M. and Paoli, A. (2020) A review of robotics and autonomous systems in the food industry: from the supply chains perspective. *Trends in Food Science & Technology* 106, 355–364.

Eckstein, D., Goellner, M., Blome, C. and Henke, M. (2015) The performance impact of supply chain agility and supply chain adaptability: the moderating effect of product complexity. *International Journal of Production Research* 53, 3028–3046.

Ellis, S.C., Henry, R.M. and Shockley, J. (2010) Buyer perceptions of supply disruption risk: a behavioral view and empirical assessment. *Journal of Operations Management* 28, 34–46.

Fartaj, S.-R., Kabir, G., Eghujovbo, V., Ali, S.M. and Paul, S.K. (2020) Modeling transportation disruptions in the supply chain of automotive parts manufacturing company. *International Journal of Production Economics* 222, 107511.

Financial Times (2020) Hauliers count losses as lorry queues grow in Kent after French port closure. Available at: https://www.ft.com/content/56125c91-cbcf-4f32-92a2-befb19222e95 (accessed 15 January 2021).

Garvey, M.D. and Carnovale, S. (2020) The rippled newsvendor: a new inventory framework for modeling supply chain risk severity in the presence of risk propagation. *International Journal of Production Economics* 228, 107752.

Garvey, M.D., Carnovale, S. and Yeniyurt, S. (2015) An analytical framework for supply network risk propagation: a Bayesian network approach. *European Journal of Operational Research* 243, 618–627.

Gaur, J., Amini, M. and Rao, A.K. (2020) The impact of supply chain disruption on the closed-loop supply chain configuration profit: a study of sourcing policies. *International Journal of Production Research* 58, 5380–5400.

Gligor, D.M. and Holcomb, M.C. (2012a) Understanding the role of logistics capabilities in achieving supply chain agility: a systematic literature review. *Supply Chain Management: An International Journal* 17, 438–453.

Gligor, D.M. and Holcomb, M.C. (2012b) Antecedents and consequences of supply chain agility: establishing the link to firm performance. *Journal of Business Logistics* 33, 295–308.

Gligor, D.M., Holcomb, M.C. and Feizabadi, J. (2016) An exploration of the strategic antecedents of firm supply chain agility: the role of a firm's orientations. *International Journal of Production Economics* 179, 24–34.

Ho, W., Zheng, T., Yildiz, H. and Talluri, S. (2015) Supply chain risk management: a literature review. *International Journal of Production Research* 53, 5031–5069.

Ivanov, D. (2020) Predicting the impacts of epidemic outbreaks on global supply chains: a simulation-based analysis on the coronavirus outbreak (COVID-19/SARS-CoV-2) case. *Transportation Research Part E: Logistics and Transportation Review* 136, 101922.

Jain, V., Kumar, S., Soni, U. and Chandra, C. (2017) Supply chain resilience: model development and empirical analysis. *International Journal of Production Research* 55, 6779–6800.

Jüttner, U., Peck, H. and Christopher, M. (2003) Supply chain risk management: outlining an agenda for future research. *International Journal of Logistics Research and Applications* 6, 197–210.

Kao, S.-C. and Wu, C. (2016) The role of creation mode and social networking mode in knowledge creation performance: mediation effect of creation process. *Information & Management* 53, 803–816.

Kumar, S. and Anbanandam, R. (2020) Impact of risk management culture on supply chain resilience: an empirical study from Indian manufacturing industry. *Proceedings of the Institution of Mechanical Engineers, Part O: Journal of Risk and Reliability* 234, 246–259.

Leat, P. and Revoredo-Giha, C. (2013) Risk and resilience in agri-food supply chains: the case of the ASDA PorkLink supply chain in Scotland. *Supply Chain Management: An International Journal* 18, 219–231.

Marucheck, A., Greis, N., Mena, C. and Cai, L. (2011) Product safety and security in the global supply chain: issues, challenges and research opportunities. *Journal of Operations Management* 29, 707–720.

Munir, M., Jajja, M.S.S., Chatha, K.A. and Farooq, S. (2020) Supply chain risk management and operational performance: the enabling role of supply chain integration. *International Journal of Production Economics* 227, 107667.

Nyamah, E.Y., Jiang, Y., Feng, Y. and Enchill, E. (2017) Agri-food supply chain performance: an empirical impact of risk. *Management Decision* 55, 872–891.

Pereira, S.C.F., Scarpin, M.R.S. and Neto, J.F. (2020) Agri-food risks and mitigations: a case study of the Brazilian mango. *Production Planning & Control: The Management of Operations* 32, 1237–1247.

Pettit, T.J., Fiksel, J. and Croxton, K.L. (2010) Ensuring supply chain resilience: development of a conceptual framework. *Journal of Business Logistics* 31, 1–21.

Pettit, T.J., Croxton, K.L. and Fiksel, J. (2013) Ensuring supply chain resilience: development and implementation of an assessment tool. *Journal of Business Logistics* 34, 46–76.

Ponomarov, S.Y. and Holcomb, M.C. (2009) Understanding the concept of supply chain resilience. *The International Journal of Logistics Management* 20, 124–143.

Ritchie, B. and Brindley, C. (2007) Supply chain risk management and performance. *International Journal of Operations & Production Management* 27, 303–322.

Sarkar, S. and Kumar, S. (2015) A behavioral experiment on inventory management with supply chain disruption. *International Journal of Production Economics* 169, 169–178.

Scheibe, K.P. and Blackhurst, J. (2018) Supply chain disruption propagation: a systemic risk and normal accident theory perspective. *International Journal of Production Research* 56, 43–59.

Sharifi, H. and Zhang, Z. (1999) A methodology for achieving agility in manufacturing organisations: an introduction. *International Journal of Production Economics* 62, 7–22.

Simchi-Levi, D., Wang, H. and Wei, Y. (2018) Increasing supply chain robustness through process flexibility and inventory. *Production and Operations Management* 27, 1476–1491.

Terazono, E. (2021) Global food prices post biggest jump in decade [online]. Available at: https://www.ft.com/content/8b5f4b4d-cbf8-4269-af2c-c94063197bbb (accessed 6 June 2021).

Urciuoli, L., Hintsa, J., Gerine Boekesteijn, E. and Mohanty, S. (2014) The resilience of energy supply chains: a multiple case study approach on oil and gas supply chains to Europe. *Supply Chain Management: An International Journal* 19, 46–63.

Yan, B., Chen, X., Cai, C. and Guan, S. (2020) Supply chain coordination of fresh agricultural products based on consumer behavior. *Computers & Operations Research* 123, 105038.

Zhang, K., Chai, Y., Yang, S.X. and Weng, D. (2011) Pre-warning analysis and application in traceability systems for food production supply chains. *Expert Systems with Applications* 38, 2500–2507.

8 This Is Not the End: Industry 4.0 and Our Future Technological Transitions

WAYNE MARTINDALE, SANDEEP JAGTAP AND LINH DUONG

The food industry is now facing an exceptional period where resilience and sustainability in supply chains have become a genuine practice, going beyond what has happened before. While the industry has always been reactive with respect to bringing innovation into agri-sourcing, manufacturing and retailing practices, there is now an absolute requirement to be ready. The global events of COVID-19 and unprojectable geopolitical conflicts, such as the Russia–Ukraine war, have placed specific pressures on food and beverage supply that are financial, social and environmentally focused. Whereas responses to globalization have been well practised since the 1990s, these current events have meant a rethinking and re-imagining of our global food system. The emergence of digitalization and the establishment of Industry 4.0 have facilitated the development of novel sectors such as vertical farms and insect proteins, which are examples of rethinking the food system. What is apparent is the need to move further and integrate processes with people and culture in the food and beverage industry, which requires people to interact with the new technologies. It is this emerging human-focused view of working with technologies and using them for community or system benefit which shows that our future is not about single-value positions in a food system where consumers require access, affordability and assurance as a triple-A standard.

Connectivity across datasets, processes and people is a core theme of this book. It is not enough to have large datasets and data-collecting capability because the critical components of analysis that filter, associate and connect them are essential in building insight and knowledge. The development of knowledge graphs that lead operators to a generic understanding of how data can be used to improve outcomes is the goal of Industry 4.0 and this book. The requirement

Corresponding author: wmartindale@lincoln.ac.uk

© CAB International 2022. *Food Industry 4.0: Unlocking Advancement Opportunities in the Food Manufacturing Sector* (W. Martindale *et al.*)
DOI: 10.1079/9781789248593.0008

to utilize existing data held in datasets compiled by national agencies is still a barrier to determining sustainable practices because so much data is latent or under-utilized. An oft-quoted principle that is attributed to several notable individuals is that we have more data than we can ever use, but what we need more is the knowledge of how to use it. This principle is important, not only for Industry 4.0 but also for the transition from Industry 4.0 to 5.0, depending on the use of semantic web technology. This data-to-knowledge transformation defines a crucial part of it in many ways. Industry 4.0 embedded digitalization into operations and practices; in doing so, it has led to the establishment of data warehouses that continuously collect information. This mass collection, storage and even ownership of data became a reality in the biological sciences at the beginning of the 21st century when the sequencing of whole genomes, including the human genome, was delivered. It highlighted the need to not only collect data sequences but also to connect gene sequences and their functions, which has become as important as collecting the data itself. It provides an important example of the value of connecting data because molecular biology will consider the idea of a single gene sequence being able to result in a specific human behaviour as something that will be very rare, if not impossible. It has been shown that the scenario of one gene and one specific outcome is extremely rare, and the same is likely to be true for any information system, including complex systems.

A big data example closer to the agri-food system is the Copernicus programme for Earth observation (EO), which has six Sentinel satellites that download over 20 terabytes of earth observation each day. Other EO programmes will result in similar download volumes so that planetary processes and the data associated with them are, quite literally, global data warehouses. These include relevant data for projecting crop production and environmental impacts such as droughts and floods, so it is highly relevant to any digital simulation or digital twin of the food system. Global and national knowledge graphs for assessing these data are being developed further so that information systems can be used at regional, local and even individual consumer scales. Some of these are demonstrated in this book. Much of the data stored is open-access and freely available. Still, knowledge graphs need to be exceptional to initially filter, associate and connect data points and domains. Without them, we are most definitely lost, and it is why the skills needed to connect the themes of resource use and industrial ecosystems are highly desirable in our industry. With this in mind, we are beginning to focus research on the establishment of structures and ontologies associated with data architectures, which can get very complex, but in simple terms it is similar to developing a library that all can use if they can fully understand the sorting methods. The language associated with this library is the operating system, which provides an ontology that can provide a means for users to interrogate data. Of course, this aligns very well with machine learning and artificial intelligence principles, both of which are emergent in the food industry as the establishment of image analysis systems gains pace in tracking and tracing products, processes and people in manufacturing environments.

Rethinking our approach to defining how the food system operates is a question put forward by each of the chapters in this book, and they demonstrate how the big data for production and consumption can be mapped or digital twinned using GIS and other platforms that provide a user interface. By re-imagining this system, we can consider production spaces thermodynamically restricted in that their boundaries control energy transfers in factories, so there is the potential to balance energy across production systems to be as close to net-zero as possible. The frameworks for such thinking are established with respect to producing foods in hostile environments and even outer space, where the basis for passive or net-zero systems is certainly not new thinking. These net-zero innovations are data-dependent, and in the case of geographic systems they develop images in map form, not rows and columns of numbers, which present a meaningful representation of production to consumption across geographic space and product lifetime. The demonstrators presented in this book still have so much to develop because new digital technologies enable tracking of operations in real time with ever finer detail concerned with processes, products and people. The presentation of static data for single time points still has much to offer, but the integration of digital technologies in manufacturing arenas can provide visualization of performance to enable immediate responses to reduce or increase resource flows when and as required. It is resilience in action, which we will see more of in our Industry 4.0 to 5.0 factories.

The big data approach has established the use of GIS, LCA and other tools in assessing food and feed production in the national and international aspects of the food system. The methods in this book utilize reported, open-access data from national agencies, and digitalization is beginning to enable the use of bespoke data from individual factories, farms or even individual consumers. For example, changing a food product formulation may result in lower energy consumption at a factory, which can be recorded within the factory as part of a carbon footprint. This product could also provide improved micro-nutrient nutrition or consumer experience resulting in reduced food waste. This could be recorded and reported by individual consumers and provide feedback to manufacturers. Data flow will become more detailed or granular and instantaneous because cultural changes are necessary, but trust in data must be robust. An example of this working across the food system is the benefits of using wearable technologies that track consumption and energy content of diet, being recognized by consumers so that they are now relatively mainstream. Although connecting these technologies to supply-chain functions will become more important, the data needed to do this is largely latent and waiting to be utilized.

Capturing large volumes of data and analysing them is achievable, as evidenced by Earth observation programmes that provide projections and determine the impact of change on global food systems. These changes are often associated with geopolitical pressures or conflict that disrupt specific trading and logistical routes such as the South China Sea, the Gulf regions and the Black Sea routes. Again, digital tools that provide a risk assessment of food commodity resources in real time using long-term remote-sensing datasets are now possible, and this book shows how this can be achieved. Providing such intelligence and insight to food manufacturers at national and regional scales is transformative in making sustainable outcomes possible. It meets the aims of the UN Sustainable Development Goals, and SDG 9 (Industry Innovation and

Infrastructure) and SDG 12 (Sustainable Consumption and Production) are both of great importance to the content of this book.

The book does not only consider how data is associated using calibrated methods already used for many standards and certifications in the food industry; it also considers the knowledge graphs that develop meaning and feedback. Environmental attributes can be assessed using LCA and assurance protocols, and the use of standards and certifications is commonplace in the food system. It is important to consider how these will change consumption and reduce food waste, for example, and digital solutions in Industry 4.0 will begin to define these. How standards will evolve in Industry 4.0 and 5.0 is still undetermined. Still, many standardizing organizations are interested in developing languages that digital technologies use to establish networks of trust and blockchains. Social attributes in these standards can be benchmarked against codes of practice, such as those of the International Labour Organization, for example, and nutritional attributes against international agencies and national regulators that determine thresholds and labelling regulations. Quality attributes of products are more difficult to define. Still, sensory assessment, consumer trials and consumer feedback trials are well established but rarely connected to manufacturing practices. It will change with the continued use of digital technologies and social media.

The description of businesses co-operating or showing dependence on each other as ecosystems is an appropriate descriptor, but applying theoretical ecology to them is not unfounded. It may prove useful because the descriptions of data, system and business interactions described in this book align with many ecological models whose simple equations, iterated many times, can project chaotic outcomes and are well demonstrated. Extending these models to markets has helped to define the food system for the demonstrators in this book. Most notably, ecological models have been used for the maximum sustainable yield projections of fisheries, determination of competition cycles for commercial risk, and simulating epidemiology of populations for enhanced biosecurity. Their use has great potential to project consumption in food systems because they enable a greater understanding of how populations respond to change. In addition, Industry 4.0 technology enables data to be efficiently captured, scaled and analysed across supply chains by established ecosystem models such as the logistic function or equation or game theory.

At the end of this book and the current study, we are asking what the future food system for food manufacturers will look like, and, if we can do this, how will they best prepared for change? Our conclusion is relatively straightforward; the viewpoint of the meal or plate is crucial, and it will change, but consumers will provide new data and require new products. Their current worldview is demanding more efficient outcomes such as net-carbon zero and zero-waste, each of them focusing attention in light of providing accessibility, affordability and assurance. It is certain that the future leaders who will shape this industry will have insight across diverse disciplines. Still, the delivery of consumer experience and values will remain as important as they ever were. With this in mind, we hope that our readers seek change for the better, meet our Sustainable Development Goals and analyse evidence more efficiently than they have ever done before.

Index

Note: Page numbers in **bold** type refer to **figures** and page numbers in *italic* type refer to *tables*

CABI – who we are and what we do

This book is published by **CABI**, an international not-for-profit organisation that improves people's lives worldwide by providing information and applying scientific expertise to solve problems in agriculture and the environment.

CABI is also a global publisher producing key scientific publications, including world renowned databases, as well as compendia, books, ebooks and full text electronic resources. We publish content in a wide range of subject areas including: agriculture and crop science / animal and veterinary sciences / ecology and conservation / environmental science / horticulture and plant sciences / human health, food science and nutrition / international development / leisure and tourism.

The profits from CABI's publishing activities enable us to work with farming communities around the world, supporting them as they battle with poor soil, invasive species and pests and diseases, to improve their livelihoods and help provide food for an ever growing population.

CABI is an international intergovernmental organisation, and we gratefully acknowledge the core financial support from our member countries (and lead agencies) including:

Discover more

To read more about CABI's work, please visit: **www.cabi.org**

Browse our books at: **www.cabi.org/bookshop**,
or explore our online products at: **www.cabi.org/publishing-products**

Interested in writing for CABI? Find our author guidelines here:
www.cabi.org/publishing-products/information-for-authors/